The Abilene Net

By

Gregg Powers

First published by AuthorHouse 04/26/04

ISBN: 1-4140-4715-0 (e-book)
ISBN: 1-4184-4226-7 (Paperback)

This book is printed on acid free paper.

Table of Contents

Preface

This book is dedicated to God who makes all things possible through his son Jesus Christ, my wonderful wife Lois, and to my first mentors in security, Troy Schumaker whose humility is exceeded only by his knowledge and Demetrios Lazarikos who boldness and combined knowledge of security and human behavior allow him to excel in all aspects of Social Engineering. Both of these gentlemen approach security the way it ought to be approached. They recognize that security is a journey, not a destination and that security is not just a science, but also an art. The humility they exhibit in dealing with their customers and others is a model for all to follow.

Foreword

This book is not designed to be a technical treatise on Internet security; others have walked this path before doing an excellent job, so there is no need to repeat their efforts. Rather, it is designed to focus on common sense as applied to the Internet and to put forth axioms related to the exchange of sensitive information. These are axioms that should be considered in the development of a layered security model. This book is written for the high level decision makers in a company who are starting to be held responsible for protecting the sensitive information of their company or government, their clients or constituents, and their employees. The axioms presented in this book are common sense axioms that make sense in any security setting and when taken together as a whole, beg the question: is the public Internet really an appropriate place to exchange sensitive information? This book will challenge traditionally held beliefs and may be quite controversial.

New security companies arise on a daily basis, many making a healthy living providing security consulting. While their services are useful, many operate from a flawed assumption. Many assume that enough security technologies and policies can fully protect a company and its information resources. This is a false assumption as we will see within this book. The dynamic nature of the environment we are trying to secure precludes any such guarantees.

Even as I wrap up this book, a new attack called the Slammer worm has been launched against the World Wide Web, slowing down access to Internet sites through an attack on database servers, hitting the financial industry hard. The Slammer worm set speed records for its ability to spread throughout the Internet. In addition, more than 8,000,000 credit card numbers were recently stolen from a credit card processing company. These are just a couple of high profile events that have recently arisen on the Internet landscape.

The first realization that a book of this type might be necessary was in discussion with executives of companies who could not clearly articulate the threats that exist on the Internet. Many take it for granted that the Internet is the de facto place to conduct business without even considering alternatives. Even if they could identify general threats, many had not taken the time to sit down and analyze the nature of these threats so that they could develop specific countermeasures. Although virtually every organization using the Internet maintains some type of security staff, rarely is it of sufficient size or skill to deal with the number and evolving complexity of threats that exist in the Internet community, nor is it sufficient to address the ever changing security landscape. Evidence of this claim can easily be validated by simply looking at the ever increasing number of incidents related to security breaches among both high and low profile companies. While annual spending rates on security related products and services will continue to increase, it is the basic philosophy, that enough technology and diligence can provide complete information security, which is in question. This book questions the common "wisdom" of using the public Internet for the exchange of sensitive information and looks at the basic security principles that should be adhered to for electronic commerce regardless of what landscape it is conducted on. Some people will read this book and may point out that it is simplistic and alarmist. I would ask you to look at the facts objectively and determine for yourself whether we are winning the war or losing the war in securing our resources and what the costs are to even make the attempt.

Introduction

Look, stop me if you've heard this one before…

Four adults are sitting on a porch in 104-degree heat in the small town of Coleman, Texas, some 53 miles from Abilene. They are engaging in as little motion as possible, drinking lemonade, watching the fan spin lazily, and occasionally playing the odd game of dominoes. The characters are a married couple and the wife's parents. At some point, the wife's father suggests they drive to Abilene to eat at a cafeteria there. The son-in-law thinks this is a crazy idea but doesn't see any need to upset the apple cart, so he goes along with it, as do the two women. They get in their Buick, which is not equipped with air conditioning, and drive through a dust storm to Abilene. They eat a mediocre lunch at the cafeteria and return to Coleman exhausted, hot, and generally unhappy with the experience. It is not until they return home that it is revealed that none of them really wanted to go to Abilene – they were just going along because they thought the others were eager to go.

Sound familiar? The Abilene Paradox is a paradox that is examined by many middle and senior management personnel during management seminars. The lesson to be learned from the Abilene Paradox of course, is that people in groups, tend to agree on courses of action that individually they know don't make sense. Examples of this "Groupthink" mentality, a term coined by Irving Janis, include the Challenger disaster, Pearl Harbor, and many others. Sometimes the paradox occurs because others take strong positions, enticing others to follow without carefully thinking through the implications of the action. At times, individuals agree with courses of action to be "socially acceptable" or to "fit in with the group" even though they may have serious reservations about the planned course of action. Often, people inadvertently transfer responsibility to other leaders assuming that they have carefully researched and thought through their recommendations and decisions. Very few people are immune to this effect and it requires people to stop and think, and then assert their objections about a given course of action, regardless of the potential consequences. At times, this can be hard to do, especially if the consequences of speaking up can be either career limiting or result in being socially outcast.

Now that we understand the basics of the Abilene paradox and "Groupthink", let me tell you about another place this paradox is being practiced on a routine basis – the Internet. Here's the way that I might sum up the Internet:

> Large numbers of organizations and individuals have elected to utilize a series of public, unsecured networks, where the population of the network is largely unknown, where any entity can obtain access to the network, where actively hostile elements exist and practice their trades, and where sensitive information about businesses, individuals, and governments is routinely exchanged.

Sound accurate? While many people may object to the way in which it is phrased, most people would agree with the assertions, namely that the Internet is a series of networks, with an unknown population complete with hostile elements, where sensitive information is routinely exchanged. If a company came in and presented the Internet to you in this manner as a part of a routine sales activity, you might laugh. You would probably indicate that you weren't interested in actually conducting business in this type of environment. Yet this is precisely what is happening.

Now, in order to be fair to the Internet Illuminati and other Internet advocates, there are mitigating circumstances to the way I summed up the Internet above. Most businesses do exchange information over encrypted channels and employ a variety of different security mechanisms and procedures to protect information. In addition, many businesses attempt to safeguard their information within their organizations, although many opt in favor of perimeter level security, leaving internal resources less protected. A mismatch continues to occur between the companies protecting information resources, which sometimes may not understand just what they are up against, believing their resources to be secure behind the corporate firewall, and the professional and criminal elements, who invest heavily (time and money) to compromise those very same information resources.

Statistics depicting the number of attacks as well as the seriousness of the attacks confirm this. We shall see, later on in this book, attempting to keep up with the hostile elements attempting to compromise your corporate or individual security is a monumental and costly task. The measures taken today, to protect information resources in storage and through transmission, lead to a false sense of security. Even though we might be able to identify companies that have not been compromised to date, I would contend that it's only a matter of time. A person who has no tickets, but routinely travels 75 MPH in a 55 MPH zone does not prove that they are safe from prosecution, only that to this point they have not been caught. Given the fact that most environments change on a daily basis, we provide our adversaries plenty of chances to catch us making a mistake.

Consider this simple security model for a moment. Suppose that we allowed access to the Pentagon by any and all individuals. We would, of course, store all information in secured areas and would carry on all conversations in locked, private rooms, talking in code. Would any of you out there suggest that this is ample security for the Pentagon? It doesn't take much to defeat this type of security because the basic enabler is in place – FPA (Full Physical Access). Access to a resource enables you, through a variety of investigative methods, to formulate and execute attacks against the security infrastructure. The attacks can be augmented through information obtained through carefully orchestrated social engineering attacks. Given the example above, how long would it take you to think of ways to compromise that security model? As it turns out, this type of example is not all that far-fetched. The Internet is access, enabling governments, companies, individuals, and others to access other resources and systems. All it takes is a single flaw or exposure in the security models of the companies using the Internet and the exposure can be exploited by criminal elements because they have the one thing they need – FEA (Full Electronic Access). Access via the public Internet is the enabler that makes all attacks possible. And since we don't control all of the access points, it is very hard to adequately protect information resources in this type of environment.

We have seen so many different viruses attacking the Internet community through so many different avenues. The Love Bug, Sir Cam, Bugbear, Slammer, Code Red, and Nimda are just a few of the viruses that propagated themselves over the Internet and they are setting records for how quickly they propagate, with some reaching worldwide penetration in a matter of hours. The Code Red virus for example, spread with alarming rapidity. CyberDefense magazine reported that the Code Red virus spread to approximately 359,000 servers in 14 hours and that at its peak, it affected more than 2,000 servers per minute. Overall more than 750,000 servers were estimated to have been compromised. They were launched through the Internet because someone had access. Let's not forget the denial of service attacks rendering electronic commerce paralyzed to conduct business, buffer overflow attacks launched against applications causing a crash of the server enabling access to the underlying operating system, directory traversal attacks exposing sensitive information, and so on. Today's information systems are complex, very complex. They are made up of more components than ever before. Is this bad? Not necessarily, but more components implies more potential vulnerabilities and subsequently many more potential attacks and attack strategies which can be employed. Often, it takes only a single vulnerability in one of the components supporting an application to launch an effective attack.

More and more disturbing trends are surfacing. Youths actively pursuing hacking careers many times devoid of conscience for the damage they cause to others, attacks on the Internet targeting critical infrastructure components, the introduction of mobile technologies extending the Internet to unwired locations, and the ever increasing philosophical gap between America and many other countries of the world. Here are some more trends that are scary.

o   More than 60% of companies have disciplined individuals for improper Internet usage

o   More than 30% have actually terminated employees for the same type of improper Internet usage o 13% of employees spend more than two hours on the Internet a day

o 125 of the Fortune 500 have battled sexual harassment claims arising from misuse of the Internet or email

The trends are leading us to a less stable, less secure world and the Internet is only one casualty of this insecurity. Is this type of network really the place that we want to be exchanging sensitive information and conducting business?

The Abilene Paradox is being relived again in the minds of many individuals and executives as they utilize the Internet as the main conduit through which to conduct electronic business and exchange sensitive information. Virtually all corporate executives believe in the value of the Internet as a transport for sharing information and for transacting business enabling them to reach millions of potential customers, yet few of them consider carefully the actual cyber risks to their companies, their customers, or their business partners.

It is this panacea of universal access to a marketplace of consumers and businesses that can overwhelm common sense. Many individuals do not consider the actively hostile elements that are aligned against them in the Internet, grouping them into a general category of security threats. Often there is too great of faith in the companies that they rely on for products and services and hoping that utilizing these products in combination with a minimal security staff, will be sufficient to protect their information and technology resources. Executives utilize the network without understanding the ever evolving threats that can be marshaled against their business, their employees, and their customers, and most importantly, the data that represents them.

Many companies employ meager security staffs, augmenting them with consultants that may not be vested with the same level of interest in securing resources as employees of the company. Many security staffs are relatively stagnant, slowly assimilating changes in technologies and closing vulnerabilities, leaving open the window of vulnerability that exists between the time an exposure is identified and the time it is patched, longer than it should be. An example of this was the recent Slammer/Sapphire worm that attacked database servers. The patch for database server had been out for more than six months, but as evidenced by the number of sites affected, many entities had not applied the patch, demonstrating that we often pay more lip service to security than sufficient action. The whole environment is like running on a treadmill trying to get to the panacea of foolproof security, but it always remains just out of reach. This is the reality of the security game – the Internet will never be an environment that is totally secure.

Security can not be regarded as a reactive activity, but must be viewed proactively. Companies need to invest in the necessary testing required to safely introduce new technologies, applications, patches, infrastructure components, etc. Executives may not give usage of the Internet a second thought, treating it as the de facto network through which to conduct business without realizing the size of the investment required to have a reasonable chance of protecting resources connected to it. They use it because they, or their staffs, perceive it as an easy conduit through which to conduct electronic business and to connect to existing and potential business partners. Unfortunately, in the desire to quickly retire manual, aging business processes with new electronic business processes, many have taken the easy road, but not the safe road. Using the Internet as a conduit for the exchange of sensitive information is not a sound idea. Before we analyze the risks associated with using this type of network for the exchange of sensitive information, let's look at some history and establish some baseline concepts and ideas that we can use to formulate our security axioms.

A Bit of History

One of the most overlooked aspects of the Internet Revolution is the history behind it. In our continual rush to evolve our computing paradigms, we continue to forget the lessons learned from earlier era computing. A short review of history is in order. Consider for a moment the computing environment in the late 1970's and early 1980's. Computing was dominated by large servers (mainframes) connected to dumb terminals. The mainframe provided its own security in that expertise was not widespread, there were limited points of entry to the system and these were not public entries, and many batch processes didn't facilitate interactive access to the system. The model had many advantages with it, namely centralized management, lower TCO (Total Cost of Ownership), and excellent performance.

With the advent of the personal computer and microcomputers, a new style of computing arose – the client / server computing model. The client / server computing model featured intelligent clients working cooperatively with servers to perform some business function. While this model was fairly readily adopted, many people, in their rush to implement client / server systems, refused to learn from the lessons of the mainframe era systems and ended up with solutions that were considerably more expensive to maintain. The new systems architecture pioneers did not carry with them the knowledge of lessons learned from history and had to relearn lessons that had already been learned with prior generation systems. When client / server systems were first developed, they were not as reliable as traditional mainframe systems nor was it as easy to find and correct problems, due to the number of components in the service delivery path.

Now, after many years of client / server computing, the rush is now back to the browser – essentially a dumb terminal using server side components to provide the large majority of the business functions. Could client /server computing have been implemented with a focus on server based business functions to start with? Absolutely, but because we didn't consider the lessons learned from prior generations and asking why things were implemented they way they were, we took the long and expensive path to distributed systems implementations.

Consider for a moment the evolution of networks. At one time, many different types of physical and logical networks existed. LAT, DECNet, LAVC, IPX/SPX, Netbios, TCP/IP, LU 6.2, X.25 were a few of the many different types of network protocols that existed. These networks did not easily interoperate and communications gateways were required to even allow the networks to exchange information in a coherent manner. When TCP/IP rose to prominence, especially in the commercial world, most adopted the standard developed back in the late 1960's and communications became much more seamless. What did we lose? In our rush to standardize on a network protocol we failed to consider the inherent protection afforded by diversification of networks. We failed to ask what the benefits of different types of technologies were. We simply assumed that everything could be connected and that would be good. While universal connectivity does have advantages, it can also have a casualty: reasonable security. While this author does not advocate a return to multiple protocols, we simply did not address the negative side effects of seamless connectivity prior to the migration.

Thus, the panacea of universal connectivity and universal access was sought without consideration for the implications. Individuals were so glad to get out of the mess that multiple protocols caused, the overhead of maintaining them, and the work required to get them to interoperate, that they rushed headlong into a whole new set of problems. Unfortunately, the new set of problems, the problem of securing resources, is much worse and can cause harm to many more people than the problems of having to work to get different protocols to interoperate.

In terms of the Internet evolution, we are at a similar maturity point as was the client / server revolution before we started heading back to the browser. In our rush to establish universal connectivity and interoperability, we have forgotten what was afforded to us through the use of partitioned systems. We have placed access and interoperability at the head of our list of requirements when protecting the information should have been at the pinnacle. We allow almost any type of entity Full Electronic Access to our communications channels. We believe in the value of cost savings achieved through technology unification, but have not considered the liabilities we incur by inadequately protecting information. We have, in seeking to simplify the administration and management of networks, opened up new holes in software, exposing even more vulnerabilities. We continue to build and rely upon technology and as we do, we reduce our participation in the process. New technology is of value, unless we assume that technology can fully protect us and we let down our guard, forsaking equally important human centric security considerations. Technology is a tool, nothing more. It can not perform miracles and it can not, in and of itself, fully protect us.

While I realize this may actually appear to be an anti-technology position, it is far from that. I as a CTO (Chief Technology Officer) believe in the value of technology and often have to evaluate new technologies that can assist in protecting our resources, but never assume that technology alone is the answer. Technology is one of the keys in a scheme to protect information resources, but it can not be applied unilaterally nor can it completely replace human diligence. Most software is not dynamic in nature. Thus, it is programmed to look for certain patterns or events and then to take a series of actions based on those events. What happens when a brand new attack materializes that was not envisioned by the developers? Often the software fails to recognize the attack and allows it to succeed. Software is developed around the abilities of its developer.

When human vision fails to anticipate and recognize an attack, the software developed by a human often has the same limitation since it is based upon human development and vision. Although security software continues to evolve, we must still augment it regularly with proactive human diligence. Be very, very afraid of vendors who claim to absolutely prevent any type of activity or market a security technology as invincible. Ensure that they are willing to back up any outlandish claims with a complete financial commitment to you and your customers in the event that their technology is compromised or your resources are compromised through it.

People might argue that we have, over time, emerged none the worse from our foray into distributed systems technology, having circled back to apply the lessons from mainframe computing to the distributed world. They might also point out that the same thing is happening to the Internet; that is eventually we will have a fully secure Internet. I would cite two arguments to offset these claims.

First and foremost, although we have now applied many of the lessons of mainframe style computing to the distributed computing paradigm, there was a very high cost paid in not learning these lessons *before* we started the migration to distributed systems. The technology changed, but the process aspect was left behind and was forced to catch up at a later time. In our initial distributed systems development and deployment, we spent large amounts of time troubleshooting performance issues, resolving bugs, dealing with security and backups, etc. We had to often deal with developing services that were already available on more mature platforms, but not available on our new distributed platform slowing the overall rate of deployment and casting us headlong into a non-standardized world. Many of these issues have now been solved, but it has taken both time and money. Today, the cost of exposing sensitive information can have significant consequences to the entity that exposes the information and the individuals about whom the information pertains to. These costs can be much higher and much more impacting to corporations and individuals than the simplistic lack of inefficiency and downtime that the move to client / server systems caused.

Secondly, with respect to the Internet, there is another difference and an important difference at that. The main difference with the Internet is that we have brilliant minds actively opposing us and these minds have the one thing they need to be successful – Full Electronic Access. We are no longer fighting only against ourselves to be successful, but we are fighting enemies who can evolve, adapt, and learn just as we do. In addition, they can generally get access to the exact same technologies that we can. While we spend time trying to get our Internet application and information exchange processes stabilized and secure, we continue to be actively opposed in our efforts.

While we will get to a more secure Internet, there will always exposures in the components that make up the Internet. Even through we have not seen this happen yet, our adversaries will eventually attack the physical infrastructure of the Internet through traditional means. While the fragmentation of the Internet limits the amount of damage that a physical attack can do to the overall infrastructure, there are still places where physical attacks could disrupt some services. Unfortunately, no technology can prevent these types of attacks.

The Interminable War

What is the war we speak of here? Well, the war of course is the constant struggle against hidden and well resourced enemies to protect your information resources while they seek to compromise those very same resources. The war is waged out over a series of networks that are spread throughout the globe, some of which are regulated, some of which are not. Cyber warfare is a reality and although we are in the embryonic stages of the war, this war is here to stay. This is not a traditional war though. In a traditional war, the battle lines are clearly drawn.

While the wealth and advanced technology of the countries such as the United States make them primary targets in this war, they are not the only targets. In this war, everyone is fair game. There is too much at stake for our enemies to let this war end gracefully. There is too much critical information that can be used in so many different ways to attack governments, harm individuals, steal funds, compromise critical infrastructures, etc. to simply let it alone. The more sensitive information that we transmit and store on the Internet or Internet accessible repositories, the bigger the target becomes. As our world becomes more hostile and more divided, the war will escalate.

If you asked most of the executives of major American companies what war they were in, you'd probably get a blank look from the majority, or perhaps even get a reference to a traditional war in their past. If they connect to the Internet and exchange sensitive information over the network, they are at war whether they know it or not. They may not realize it, they may not act upon it, or they may dismiss it as a trivial threat. There are people on the Internet that want to seriously harm your company, your employees, your stock price, and your customers. This is a war; it can not be dismissed out of hand and it requires considerable preparation and a comprehensive strategy to be able to have any chance of defending against the onslaught of hostile elements residing on the Internet.

The problem with this war is that there is no direct way to win it. In a traditional war, the war is over when your enemies surrender or you eliminate them. From the perspective of the United States and other legitimate entities, this is a purely defensive war and most organizations don't seek to eliminate their enemies, but rather seek to protect their resources. Although traditional law enforcement entities may be seen as the Calvary coming over the hill, many law enforcement agencies are just now beginning to evolve their organizations to deal with this type of criminal activity. Even when they enter into the war, they can not win it. They can only assist in waging it.

Organizations can struggle with how much to invest in security since there is no prescribed amount of dollars or resources which can be expended that will guarantee protection of an organization's information. Security preparedness is similar to a high-jump. When a minimal amount of resources and processes are invested to secure information, the bar is set relatively low, protecting only against the most novice threats. As more security is layered on, the bar is raised so that successively less and less entities are able to breach the security. This metaphor raises an interesting question: Can the bar be raised high enough to prevent all security breaches? The answer is likely NO. As we read through this book, we will see why not.

There are two main factors that prevent us from completely protecting resources. The first is lack of vision and the second is a detailed understanding of the issues. Lack of vision causes us problems because our ability to develop countermeasures is squarely dependent on our ability to envision a threat. Employing others, such as professional security firms, can help us to envision more threats, but even they will not be perfect. The depth of our ability to envision, and predict the nature of threats and how they might be carried out, is limited. There are individuals that are better and worse at this task, but there are NO perfects. If someone claims that they can guarantee that they can protect you and your resources, run, don't walk from that level of arrogance.

Our ability to envision threats depend on our knowledge, our values, and our experience. Knowledge is helpful for obvious reasons. Experience can form the basis of knowledge, replacing theoretical knowledge with empirical knowledge. Experience should be the most sought after of commodities, which is valuable as both a learning tool and a platform from which to envision a wider array of possibilities. Unfortunately, experience can help us to understand what might evolve in the future, but there is no guarantee it is the basis of envisioning all possible attacks, especially in a dynamically changing environment.

Our values also come into play when, because of a particular set of beliefs, we can not envision people taking a specific course of action. An example of this occurred on September 11th 2001, when commercial planes from United States companies were flown into the World Trade Center buildings and the Pentagon. Why did the people on the plane allow this? It was most likely not out of fear, but because their value system handicapped them from considering the eventual outcome a possibility. Passengers on the planes that crashed into the World Trade Centers and the Pentagon most likely assumed that this was a routine hijacking that would result in an endgame of them landing in another county, negotiations taking place, and eventual release. This line of thinking probably led them to accept what was taking place because they assumed they knew what the endgame was. The opposite happened on the flight bound for the White House. When the passengers found out what happened to the other planes and began to believe that their endgame was to die crashing into a building, they accepted the situation, and heroically rushed the cockpit preventing the deaths of 100's more.

We misunderstand issues because we don't realize that we are in a war. We take limited numbers of precautionary measures and meagerly invest in a small security staff in order to protect our resources. Unfortunately, we continue to see over and over that this is simply not sufficient. We operate from the perspective that technology can solve our problems and so we rely on technology only to be disappointed over and over. We must all evolve our understanding, beliefs, and knowledge of security issues in order to attempt to effectively combat our adversaries.

Consider now, some of the platforms issues which need to be addressed in order to raise the security bar to a level where information is reasonably protected in a moderately complex application. The rate at which the environment continues to change is accelerating. The web-based applications of today require a large number of components to operate and then of course there are the communications between those components. Consider this partial litany of components often used, either directly, or indirectly, in a typical web-based application.

Component	Operating System	User Info	Functional Software	Business Data	File System
Application Server	✓	✓	✓	✓	✓
Database Server	✓	✓	✓	✓	✓
Directory Server	✓	✓	✓	✓	✓
Web Server	✓	✓	✓	✓	✓
E-Mail Server	✓	✓	✓	✓	✓
Router	✓	✓	✓	✓	✓
Firewall	✓	✓	✓	✓	✓
Switch	✓	✓	✓	✓	✓
Gateway	✓	✓	✓	✓	✓

These are just representative components in some typical web-based applications. While some of the above components do not have customer data resident on them, almost all of them have some type of administrative access through which system parameters are set or configured. Depending on the complexity of an individual application, there can be many more components that participate in the delivery of an application which have not been listed here. Each one of these components can be attacked on multiple levels. For example a database server can be attacked through the operating system, the database management software, the file system, during transmission of information, etc. Vulnerabilities within any one of these can cause problems.

Trying to keep a constantly changing environment secure is highly problematic, if it is even feasible at all. There are too many places in the complex computer systems of today into which humans can introduce errors. It will never be possible to fully protect systems and to maintain that protection over time. As an organization grows and it deploys more technologies, there is an increasing level of commitment required to keep an organization at the same security posture. The systems we develop today may be technically simple compared to the systems of tomorrow. Consider where we have come from and you will see the dramatic increase in complexity of the applications that we currently develop when compared to the applications of the 1970's, 1980's, and even the 1990's. Evidence of this is easy to acquire by researching the number of lines of code in successive versions of popular operating systems.

Cyber warfare is almost invisible, except to its victims. Countries participating in cyber warfare do not declare war in the traditional sense. The battle is somewhat one sided. You have an entity that is constantly attacking and an entity that is constantly defending. When attacks are launched from countries sponsoring this type of activity, it can be extremely difficult to catch or prosecute them, even if the launch point of the attack can be ascertained. In a recent security seminar that I attended, one of the guest speakers discussed the overall surveillance of packets that traversed the Internet. They indicated that in 1995, only 20% of the packets that traversed the Internet were captured and inspected. In 2001, this number rose to 89% of the packets that traversed the Internet.

Speaking from personal experience, there have been a considerable number of attacks on my own home system. I have virtually all TCP

(Transmission Control Protocol) ports blocked to my home system and yet there are routine attempts from a variety of networks to break into my system. I am a target and yet I have very limited information within my home system. This tends to be the rule and not the exception. In addition, a recent report published by Symantec estimates that the average company is attacked 30 times per week up from 25 the previous year. Many of these attacks are not designed to be attacks that compromise systems, but rather vulnerability reconnoitering, seeking to expose any known holes in existing products.

We might conclude, in general, that our defenses are working, but this is a hasty conclusion. A more reasonable conclusion is that to this point, hostile elements have not found a way to breach every company's resources. Time however is on their side. They know that even if they can not break into a company's systems today, there will always be a new opportunity in the future. Even though large numbers of attacks may be thwarted, a single exposure can be very lucrative. Within the past month, a credit card processors database was attacked exposing ~ 8,000,000 credit card numbers. The payoff can be so lucrative that hostile elements can afford to be patient and attack a large number of sites over a long period of time. Hacking no longer requires personal involvement. Often times the attacks can be launched programmatically with additional attention required only if a probing initiative succeeds.

Once again, we see the very nature of a public network coming into focus – a place where individuals can hone their skills of breaking into systems, intercepting information packets, and trying to decrypt encrypted packets. They can practice these skills in virtual anonymity. What drives this type of behavior? The potential rewards of course. They can intercept credit card numbers, capture bank account numbers, PINs (Personal Information Numbers), social security numbers, birth dates, addresses, and other types of sensitive information. With this information they can steal identities, purchase items, impersonate individuals, transfer funds, request credit cards, etc. While the damage that can be done to an individual is generally limited to that individual (there are cases where this is not true), businesses exchanging information have much more to lose. They collect and store large amounts of data, often on millions of people. Not only are their assets more extensive, but they also exchange even more sensitive information than most individuals. Stock information, funds exchanges, quarterly results, procurement, intellectual property, health information, credit card numbers of individuals, personal information, etc. are all examples of information that routinely flows across open networks.

The really insidious aspect of this covert inspection is that often times, companies have no idea that sensitive information has been collected or compromised until it is too late to respond to them. Hostile elements are looking for ways into systems and often this probing can be done without actually launching an attack. To this end cyber warfare is similar to traditional warfare in that an attacker scouts out his enemies' weaknesses and then attacks.

Intercepting and stealing information is not the only way to harm a company. One of the common attack types is called the DOS (Denial-Of-Service) attack. DOS attacks don't attempt to gain access to sensitive information, but they do seek to prevent others from doing business. DOS attacks flood selected targets with packets that can overwhelm host systems beyond their capacity to respond. While this can do limited damage to a system owned by an individual, consider the damage that can occur when the same type of attack is levied against a corporate system. All of a sudden revenue streams can be impaired or eliminated for a period of time. An effective DOS attack can literally cost millions of dollars per hour. Many consumers will turn to alternate sources of products or services rather than wait through the poor performance or lack of service from a given vendor. But wait, it gets worse.

There are a number of key services and systems that power the Internet. What if these services and servers were put under the same type of attack? The results could be that users could not resolve URL's (Universal Resource Locators such as www.website.com) through the DNS (Domain Name System) resulting in the inability to connect to web sites or worse. And while we tend to think of isolated systems attacking individuals systems, envision 100 or 1000 different systems attacking hundreds of servers, through different networks, in unison. The reality of this is that there is simply is no way to guarantee a protected network where the network population is a mystery and where the attacks can come from a large number of sources. There are simply no international treaties or agreements with all countries and individuals connected to the Internet to track down or prosecute the originators of these attacks, especially in rogue states which may be more interested in attacking the Internet infrastructure than building it up.

In early February 2000, a denial of service attack was launched against the Yahoo web site. Research on the attack after it occurred revealed some interesting facts about the nature of the attack. First, it was a coordinated and distributed attack meaning that multiple systems connected to the Internet coordinated to attack the Yahoo site. This is especially noteworthy since a single PC can do limited damage often due to limitations of the connection speed to the Internet. The denial of service attack originated from 50 different addresses which levied a total of 1 Gigabyte of requests per second against Yahoo's multiple web servers. While the exact cost of the attack was not known, analysts estimated the cost to be well into the millions of dollars. As dramatic as this might be, imagine a much larger number of nodes doing the same type of thing. While there are ways to thwart this particular type of attack, this is only one such type of attack.

The war that we have entered is here to stay. We must begin seeking ways to protect ourselves in a manner that is more secure. When we wage this war over a network that they have access to, we wage war according to their rules, on a battlefield that we don't control. Considering alternatives, rather than trying to secure the un-securable, only makes sense.

The Basic Challenges

There are a number of different challenges that exist when accepting the Internet as the default place to conduct commerce and to exchange sensitive information. Most entities using the Internet are ill prepared to engage in the war that they have inadvertently entered into. Executives in large corporations are often hard pressed to explain why the Internet, a collection of networks containing unknown users, known criminal elements, terrorists, fanatics, rogue governments, as well as a collection of questionable business pursuits such as pornography, human slavery, and the like is a proper place to conduct electronic business. The leap to the Internet was likely made since the infrastructure was in already in place; that it was easier to adapt the Internet to business than develop a new set of networks that supported the exchange of sensitive information and housing a known population. It is convenient, cheap, and easy to get connected to the Internet.

It is difficult to make the hard choice of not using the Internet for the exchange of sensitive information by selecting convenience over safety. To this point, few corporate officers have been held legally liable for corporate profitability, corporate security, or the damage that could be done by transmitting sensitive information over a public network without sufficient security measures in place. We have to consider that there are alternatives that we could use to protect this information. This current lack of accountability will likely change over time and executives will be held directly responsible for direct and indirect damages occurring from our negligence in protecting information.

In the future, there may very well be a credible basis for a lawsuit against company officers when loss occurs to company resources, employees or customers due to insufficient safeguards protecting sensitive information transmission and storage, especially when alternatives are available, but not used. Even as we speak, new legislation is being considered to force the declaration of attacks on a company's resources. California is one of the first states to pass legislation that requires companies which are hacked to provide information about the nature of the breach to the victims, to the media, and to others about the nature of the attacks. We continue to see an accelerating number of attacks against entities operating on the Internet, and as bad as it is now, it will only get worse in the future.

We must also accept the political reality of this world. Bringing politics into a technology based discussion is important here because the reality is that the United States and some of its allies have different political ideals than many of the other countries of the world. This division in values can lead to a constant friction between different countries. When there is a large delta in the Standard of Living in respective countries, this only serves to magnify the differences between them. This is important because one of the ways to attack a country is through its constituent companies, directly affecting its economic infrastructure. With respect to the United States, attacking American companies achieves two different objectives valuable in the crusades carried out against the country. First and foremost, the opportunity to attack the economic infrastructure of the United States can be achieved by attacking United States companies. This is exactly what happened on September 11, 2001. The terrorists used the assets of two American companies to attack different pieces of critical infrastructure – the Pentagon (military), the World Trade Centers (economic), and the White House (government). Although the attack on the White House failed, it demonstrates the terrorist's ability to use our own tools and assets against us. Resourcefulness is not something that our adversaries are lacking.

Even to this day, we as Americans are haunted by the events of September 11th and we have become somewhat economically stagnant. Although we are starting to recover from the impact of those attacks, there will likely be other attacks and only time will tell how well we are mentally prepared for those attacks.

The second thing that entities attacking American companies gain from the attacks is the feeling of vulnerability by its citizens. As Americans, we have lived with an air of invulnerability for many years. Save our internal wars and the initial attacks of World War II by the Japanese, we have never fought a major war on our own soil. Much of the rest of the world is on constant guard against attacks or terrorist activities and now we must be as well. Being able to compromise the systems of American companies adds to the feeling of vulnerability experienced by Americans and causes people to think twice about the safety of their information. Often our enemies will use our own tools against us.

One of these tools that they will continue to use against us is the Internet, especially if we are willing to utilize the Internet as a place to transmit sensitive information. They will also not hesitate to use, if possible, our software, our systems, and our people to attack us with. We are prime targets for those entities throughout the world who do not share our philosophies, our values, and our dreams. The United States and many other countries have been greatly blessed by God to the point where we have many different freedoms (political, religious, and economic). We must face the fact that we are a target and respond appropriately, but never assume that we as a people are superior or that our technology is so superior that we don't have to give threats from external entities a second thought. We must evaluate our position in the world considering risks that exist from entities outside the United States and within the United States.

Risk management is a subjective process largely because of the human factor in evaluating risk. In many cases, individuals will not agree on the level of risk for some event and may not even agree on the consequences of the event being realized. Internet usage, for the most part, is taken for granted, bypassing the normal evaluation of risk that should be invoked when using the Internet. There are so many different threats and so many different entities after our sensitive information that avoiding the risk analysis is very dangerous.

The liabilities that can be incurred when we transmit sensitive information over an open, public network can be considerable, yet this often does not factor into the decision as to whether or not to use the Internet. The Internet is a battlefield and the forces of law and lawlessness are engaged in the battle. All companies should step back and candidly evaluate alternatives to using the Internet or at the very least evaluate and determine the level of risk, augmenting security resources to reduce the risk. If it is determined that the risk incurred is too great, or requires too high a cost to minimize the risk when using the Internet, alternatives must be sought.

Cyber warfare is here and is being practiced aggressively by our enemies. They continue to devote themselves to honing their skills and to acquiring the necessary tools and expertise to succeed in their attacks. Our enemies are preparing for war. Few of us regard ourselves as engaged in a war, yet that is exactly the case. We may become complacent because our companies haven't been attacked yet. Thus, we assume that our security measures are sufficient. Unfortunately, you can not use a clean history to testify to the validity of your security model. Just ask those who had a clean history before they were attacked and compromised.

The Internet: Target One

Willie Sutton was once asked why he robbed banks to which he responded, "Because that's where the money is!". The Internet is beginning to exhibit the same attributes that banks do, not with money, but with information. However, in the Information Age, information is money. As more and more sensitive information flows across the Internet, the Internet becomes a bigger and bigger target for criminal activities. The potential payoff continues to increase as we attach more information either directly or indirectly to the Internet. By raising the potential payoff, we entice hostile elements to attempt even bolder and more devastating attacks.

New initiatives are seeking to build even more trusted relationships. For example the Liberty Alliance project is designed to implement a federated security model which can allow an entity in one organization to access information and resources in a different organization. This is effectively a trusted relationship. If a user can convince one site of their authenticity, they can then access other resources and information that belong to trusted business partners according to the level of authorization that each user has. If one organization can be compromised, the potential exists to compromise others through the security of the source organization. In today's environment, this is protected because each site has a different login and password so if a single logon and password are stolen the damage is limited to that site. Although all security credentials are not stored in one organization, this violates a best security principle, namely that security should be implemented in layers. The federated model uses more of a perimeter security model in that once the frontline security of the local organization has been bypassed, access to other resources are possible without passing through successive layers of security. This is yet another case where in trying to knock down barriers to electronic commerce, we have deprecated security to a secondary consideration. (This author acknowledges however, that we are creatures of habit and often use the same login and password to multiple sites).

Just as IP (Internet Protocol) has become the network protocol standard, so other types of electronic commerce and security mechanisms are being standardized. Standardization can be a good thing so long as the products adhering to the standard are secure.

Open up a vulnerability in the implementation of the standard and often you have a recipe for disaster because many systems can be compromised using the same vulnerability. In addition, if the standard is widely implemented, it can take a long period of time for the fixes to be widely implemented. This is the same premise that worms use to invade many systems – exploitation of a common vulnerability.

As a further testament to high profile targets drawing the majority of the attention, a recent study by Symantec revealed that most of the recent attacks were launched against companies with more than 5,000 employees, the next most against companies with 1,000 – 4.999 employees, and so on down the line. Logically, this makes sense because the larger companies traditionally have more data, more applications, and more sensitive information than smaller companies. Company types that drew the largest number of attacks were power/energy, non-profit organizations, telecommunications, high tech, and financial services respectively. Notice that high on the list of priority targets were the power and energy companies. This is very likely criminal elements seeking to find ways to attack one type of critical infrastructure. Consider the impact that a power outage over a large area would have. Businesses, schools, homes, and governments would all be without power and their ability to operate severely impaired.

One of the other issues with repositories of sensitive information is that they are normally stored in an unencrypted manner. Many enterprise wide repositories allow widespread access by internal users for read access. They tend to rely on perimeter security to protect the data stored in their corporate repositories and in many cases do not exercise the same level of security on the database feeling that the network security is protection enough. This is a risky proposition since once the front line network security is breached access to backend databases often follows. This concept demonstrates the first of our key security axioms: **Each layer of security should be implemented as though there are no other security layers.** Too often in our attempts to protect information we assume the existence of robust alternate layers of security and thereby we allow the security we are implementing to be implemented in a less than optimal manner. By applying this axiom to our security implementations, we seek to build truly layered security models so that a given security layer does not have to rely on other security layers to protect the things it is designed to protect.

Centralized information isn't the only high profile target, although it potentially has the greatest potential for payoff. Infrastructure components such as operating systems and databases are attacked as well. Why? It is because these are access mechanisms to get to the information that they protect. Although product security might be considered weaker in many desktop based operating systems and products, the real reason that desktop technologies continue to get exploited is because they are so pervasive in the market. Think about it from the perspective of a malicious hacker. Where do you have the opportunity to obtain the best value for the investment of your hacking time and dollar? Attacking weaknesses in a reasonably small market share operating system or database system just doesn't pay off, unless you are seeking some access to information within a system protected by that small market share operating system or database system. In general, to cause the most damage or to realize the largest financial gain, you want to attack operating systems that are widely deployed. Fear not though, as other open source operating systems and proprietary software components rise in popularity, they will begin to garner attention from malicious elements.

Information Half Life

Information that moves over the Internet or stored in a repository that is attached to the Internet does not have to be intercepted and read in real time; it only needs to be captured in real time. What does this mean? It means that with some of today's encryption algorithms and longer keys, it is not possible to crack information in real time. This often leads to the mistaken conclusion that so long as it can not be read in transit, it will be safe. If care in deploying an application is not taken, there is a chance that the information will accidentally be transmitted in clear text so no decryption effort is required. Information that flows over the Internet may be captured tomorrow and decrypted today, tomorrow, next year, or never.

This is especially important because some of our adversaries have shown patience in attacking us. They did so on September 11, 2001 when they tried to commandeer even more planes then they actually got their hands on. They did this to take maximum effect of the vulnerability. If they had commandeered a single plane and crashed it into a building, we would have likely undertaken measures to prevent it from happening again and the damage from the vulnerability that they discovered would be limited. The same is true for information in that even if information is compromised they may not use it right away, preferring to defer to a time when they can obtain the maximum value from the vulnerability. This is especially true if the information was taken but the theft was not detected.

Thus, I introduce the concept of *Information Half Life.* Information Half Life is a term designed to represent the usefulness of sensitive information with respect to time. Here's a quick, and rather dramatic example of the half life concept. Suppose for a moment that you wanted to do something crazy like blow up a plane that a certain individual is traveling on. Something akin to this happened in 1987 when a former PSA employee who was reported as irked at his boss, boarded the same flight as his boss. Once the plane had reached 29,000 feet over San Luis Obispo, CA, he calmly proceeded to shoot both pilots, resulting in the death of all aboard.

Now back to our example. Let's further assume for a moment that the reservation was made over the Internet, by the individual using an encrypted transmission. While the person seeking to blow up the plane may not be able to read the information while it is traversing the network (remember that it's encrypted), all of the packets can be intercepted and stored by the person wanting to blow up the plane. Once they have been stored, computers can immediately work on trying to break the key using brute force attacks (if the key length is short enough), dictionary attacks, selective key attacks, etc. If the code is broken, it can be discovered that this individual will be flying on an identified flight exactly four months from today.

This now brings us to the concept of Information Half Life. Although the information is rendered useless four months from the time it was transmitted (because the information will have expired after four months since the flight will have already occurred) the number of options that can be levied against the plane carrying the passenger, will decrease over time. If you have four months to plan to deal with the passenger on the plane, you have a wide array of options, if you have two months before the passenger flies, you have fewer options, if you have one month, even fewer options, and if you have two minutes, you have virtually no options.

Although we don't want to get into a lengthy discussion of how to affect an airplane with four months notice, just consider the number of things that you could do with that much time. If you are creative you can come up with a large number of ways that you might be able to achieve your objective. Thus, the longer you have before the information expires, the larger number of options you have to use the information.

What can be critical here is to consider the amount of time required to break the keys or read the data (if an encryption error was made). If it takes more than four months, the information has expired, but if it takes less, some time will be left where the information can be used to plan an attack against the flight – the less time left, the less useful the information will be. This same concept applies to virtually all sensitive information.

If an error is made in the deployment of an application and the information is transmitted over an unencrypted channel, then no time will be required to get access to the information. It is also possible in some scenarios, to redirect an encrypted HTTP session to an unencrypted one. Since applications are developed using multiple components, it is also possible that access to the data could be obtained before it is encrypted. Some organizations focus only on encrypting communications between an end user and the web server though which they conduct business. If a method to get access to the data before it is encrypted is discovered, the encryption at the boundary may have limited value. In all, there are multiple ways to get access to unencrypted data even though it is encrypted between different points within the application flow. Information can often be captured and interrogated programmatically.

By way of a secondary example, consider credit cards. If I intercept an encrypted credit card number which has an expiration date two years in advance, and it takes me two months to crack the code, I have the potential to do much more damage than if I intercept an encrypted credit card number which has an expiration date of tomorrow, even if I am able to crack it immediately. This is a key concept since some information can have a very long half life. Information such as the location of nuclear missile silos, social security numbers, location and connections of critical infrastructures (electricity, gas lines, etc) all have a very long information half life, because they are static and don't often change. Thus, if this type of sensitive information flows over the Internet in an encrypted manner and is intercepted and captured, it is very possible that it will eventually be cracked and still be useful information. There are of course, exceptions to the strict adherence to the half-life concept as applied to information, but in general the concept holds for information.

This brings us to another of our security axioms: **Just because information is protected at a point in time, does not mean that it is always protected** and its corollary **Even though information may not be used against you today does not mean that it will not be used against you in the future.**

This can cause problems because people don't always consider the temporal dimension when looking at the transmission of data. If a sensitive piece of information traverses the network successfully, it does not imply that we are safe. The information may still be used against us in the future.

If the information being transmitted has a relatively short half-life, it may be possible to transmit the information over an open network, in an encrypted manner without significant risk. A risk analysis can be performed to determine the extent of the liability in the event that the sensitive information transmitted over an open network is compromised. The risk analysis may lead to the conclusion that the compromising of that information results in an acceptable liability. In this situation, the information to be transmitted would have a low liability value. If the information that was transmitted would result in a large liability, the information transmitted would have a high liability value.

When attempting to calculate the information liability, it is important to consider the problem from each entity's viewpoint. For example, perhaps no hard liability for a company would exist if the customer database was compromised and the credit card numbers of your customers were stolen. However, your customers could sustain hard liabilities and the company might lose business over the leaking of the information. As it becomes more and more visible to the end consumer, which organizations are protecting their data and which organizations are not protecting their data, business will shift. I know that I personally will move away from vendors who have not demonstrated due diligence in protecting my data. How will I as a consumer know this? Unfortunately, I don't have visibility into many of the companies that I do business with so the only thing I can base it on will be the press surrounding security incidents. Unfortunately for organizations, this visibility, through new laws, is improving to the consumer. In very selective cases, I will utilize a single credit card for a dedicated vendor. The problem is not necessarily that the information can not be transmitted over an open network, but rather there is too little thought going into the decisions as to what gets transmitted over an open network and what does not.

Longer keys reduce the overall chance that the information will be compromised, but do not guarantee that it won't be compromised. If the information to be transmitted has a significantly long half-life or if the potential liability of exposure is high, transmitting this type of information over a closed or private network is likely a sound business decision. While it would be feasible to perform an analysis of all data to be transmitted, this could take considerable time and thought. If there is any question, using a private network would minimize the chances of its being intercepted and captured, possibly for future use. This leads us to another of our key security axioms: **Transmitting sensitive information, with a high liability value and a long half life, over an unsecured network, is inherently risky.**

43

It is not only the previously sent data that is at risk, but also future data can also be at risk. Once a key has been cracked, the same key may be used to decipher additional messages that are sent at a later date unless some sort of dynamic key scheme is leveraged in the transmission. How would we know that we have been compromised? We might not, until the attack is launched. As discussed before, this can be extremely damaging if an entity believes that it is safely sending encrypted messages across the Internet, but a hostile entity is capturing and decrypting those messages.

The Constantly Changing Face of War

The war itself is a dynamic one. In the computing environments that we maintain, there are constant changes in applications, infrastructure software, hardware, security techniques, configurations, algorithms, web pages, data, knowledge, etc. The overall environment is very dynamic, so even if we find a way to completely and utterly secure a resource today, tomorrow it will change. Almost all changes to components, or applications, or security within a computing environment have the *potential* to expose new vulnerabilities. Our businesses require these changes; we don't do them for superfluous reasons. There are so many changes that occur in the infrastructure to support the dynamic aspects of the business, that it is virtually impossible to keep an environment secure. If the components that made up an infrastructure were static, the job of securing them would be a much simpler task, but with the constantly changing landscape of technology, it is virtually impossible.

Listed in Appendix A, is a list of software security vulnerabilities discovered over a period of 21 days late in 2002. Looking at the chart, we can see that there was an average of two and a half new vulnerabilities discovered each day and these are the ones being reported. Remember that hackers also discover new vulnerabilities and they don't really report these to a collection and publishing entity, but rather they use them for their own ends. The yearly average is closer to seven per day.

Another set of changes that occurs quite often is the normal configuration changes that occur in network devices and systems. While diligent administrators (network and systems) will attempt to protect the systems they are responsible for, every one makes mistakes from time to time. All it takes is the miscoding of an access control list, a change in the firewall traffic rules, etc. and vulnerability has been created. In October of 2002, a change in a router affected about 20% of Worldcom's IP network. Since it only takes a single flaw or error to enable access to corporate resources, we must be more cautious of the routine changes in our environments. Finding out about the mistake can happen too late, only after the information resources within the corporate network have been compromised.

This is one of the reasons that entities will not insure or guarantee that an environment will stay secure. If a consulting organization comes into your company and offers to secure your environment, what they are offering to do is to improve your security. If you have a security consulting company that insists it will guarantee protection of your information resources, get that in writing. There is so much work to do to completely secure an environment that it takes a lot of investigation and a long time. Even when they are thorough in attempting to secure the environment, the environment has only been secured according to the "best information" available at that time and for that computing infrastructure at that point in time. This means that if changes are introduced after the security company leaves, the environment that was secure when they left, may no longer be. In addition, changes in the environment can quickly invalidate some of the recommendations of the security company, especially if new technologies are introduced.

You can easily see how fast the war is changing. Even when a company thinks that they have a relatively secure environment, holes are almost always found at a later date. In addition, it just takes something as small as a configuration change or a new patch of the operating system to open up new holes. If you take the time to map out all of the components in a typical application, you will find there are quite a few components where vulnerabilities can be exploited.

Adding on to this dynamic landscape are newly developed types of attacks that occur on a regular basis. Security information is transient. Thus, if I know everything possible thing about security today, tomorrow the information is dated. If I went away for six months to a remote island and then returned, even less of my knowledge is still valuable. This is not meant to cast a pall on the knowledge of security experts, rather it is makes the point that security can not be viewed as a static activity. Learning must be continuous and proactive.

Finally new initiatives continue to evolve within the Internet space. These new initiatives also change the landscape of the Internet to a point where new vulnerabilities are exposed because of the changes in the infrastructure. Such an example is the identity management proposed by the Liberty Alliance Project or the Web Services standards which are only now beginning to focus seriously on security. New protocols are routinely being developed and implemented that force us to risk opening up additional exposures. An example of this type of protocols is the IPP (Internet Printing Protocol). This type of technology is designed to bypass typical firewall configurations.

A Sample Exercise

Take a moment to perform a quick mental exercise. Imagine for a moment that you have been assigned to secure a cornfield from all potential threats to the crops. We use this example because it is one to which everyone can relate. In our exercise, the cornfield is analogous to the Internet, because there are many different points of access. If you like, write down what you would do to secure the cornfield and protect the crops, being as thorough as possible. Now, envision yourself as being tasked with determining the ways you could destroy, damage, or steal the corn from the cornfield. Once again, write down what you would do to destroy or steal the corn from the cornfield. For most people, the second list will be longer than the first and there will be things on the second list that you had not thought of in the first list. Here's my admittedly hasty list.

Ways to Protect the Cornfield	Ways to Attack the Cornfield
Build a fence around it	Tunnel in from underneath and alter the acidity of the soil
Station guards around it	Spray poisons or defoliant from the air
Apply pesticides to the field	Introduce harmful insects or rodents into the fields
Setup cameras to monitor the field	Go and steal or cut down the crops at night
Place a sunlight transmitting, permeable dome over the entire field	Drop a bomb on the field
Disguise the cornfield as something else	Poison the water supply which irrigates the field
	Introduce a virus into the corn
	Launch a missile at the field to blow it up
	Launch a missile over the field to deploy a biological agent
	Set fire to the field using an accelerant
	Alter the harvesting device to destroy the crops at harvest
	Chemically alter the fertilizer

I believe that if you try the same exercise, you will probably be able to come up with many ways to protect the field and even more ways to destroy the field. It becomes clear from this exercise that it is easier to destroy than to build, to break into than secure. This has been the case for all of man's history on earth and not just for information resources. Also, you will also find that if presented with a list of security remedies (e.g. the ways to secure the cornfield in this case), it will not take you long to be able to devise a method to circumvent the protection or security. This brings us to one of our security axioms: **The first step in defeating a security scheme is to know of its existence and its nature, therefore obscuring a security scheme provides some inherent protection.**

How does this affect technology and social engineering? If I want to attack an entity, especially from anonymity, I can generally begin an investigative probing process to determine what types of technologies are in use by the company I want to attack. Once I have obtained technology specific information (either through a social engineering attack such as a survey or through investigative probing), I can begin to launch attacks at known vulnerabilities. Some may be successful and some may not be, but I as an attacker, can begin to develop a profile of my target understanding to what level systems have been protected and how they have been protected. Hackers who have been caught in the past have had, in their possession, key information about personnel inside of an organization (perhaps to launch a social attack), technologies used, applications used, areas to probe, network information, etc. Our adversaries can be very methodical in their efforts to compromise our systems. These techniques are not very different from the techniques that scientists employ when they can not validate something by direct observation.

In order to effectively defend a resource, the people defending the resource must be able to completely protect the resource by envisioning every possible attack (or perhaps class of attack depending on the subtle differences within a class) and developing a countermeasure against that type of attack. The countermeasure must either protect the resource completely or it must establish a prohibitively high cost to breach the security protecting the resource. Now to the underlying problem: an attacker attempting to breach a specific secure resource need only find one way to be able to compromise the security whereas the defender must consider all possible avenues of attack and develop sufficient security measures against them. This exercise reveals two more of our key security axioms: **It will never be possible to fully envision and counter all possible threats against a resource in an uncontrolled environment. Therefore, in order to have a reasonable chance of protecting a resource, the sources of threats must be reduced to a manageable level** and its corollary **It is easier to attack a resource than to defend a resource since an attacker need only find one open avenue of attack whereas the defender needs to anticipate and close all potential avenues of attack**. There is no current foolproof method of protecting information resources.

Understanding the Threats

In Bruce Schneier's book <u>Secrets and Lies</u>, he discusses attack trees as a method of being able to work through attack scenarios so as to be able to devise defenses against them. While the concept is sound enough and can be effective, especially for simple items which have a limited number of attack strategies, I believe it relies on a false assumption. That false assumption is that an individual or a group of individuals can effectively envision and thwart all possible attack scenarios, especially when dealing with complex resources exposing large numbers of attack scenarios. While it is possible to develop defenses against expected or known attack scenarios, it is not possible to envision all possible attack scenarios, especially for complex items like software. Developing an attack tree for a standardized file transfer program is reasonable activity, developing an attack tree for an application which may use 20 or more individual components operating cooperatively is another matter.

In the early 1990's vendors were accused of putting undocumented features in its software and operating systems. If I, as an end user of this operating system, attempt to develop an attack tree for the operating system, I can only do this for the things that I know about (and it still is virtually impossible to envision all possible threats). There may be a whole set of attacks that can be launched through undocumented features of software or hardware of which I may have

51

no knowledge. Thus, an attack tree is a good concept, but the more complex the target, the less successful attack trees will be in identifying and developing countermeasures to all possible attack routes.

There are several possibilities that explain the large number of vulnerabilities that continue to be exposed in software. The first is that individuals that develop software are willingly ignoring known attack scenarios, which is clearly unlikely. Most software developers are conscientious in nature and try their best to develop secure software. Knowledge of how to do this may be lacking, but the intent is generally there.

Second is the possibility that most software developers do not design with security as a prime consideration and that they simply do not have knowledge of "secure coding principles" to be applied during their development activities. This is one of the more likely scenarios. We can see, from an empirical viewpoint, as demonstrated through a continued increase in vulnerabilities, that software has many vulnerabilities. Most developers simply do not have a broad enough background in order to develop truly secure software. Developing secure software embodies the "weakest link in the chain" principle. That is, a secure software development process is only as good as the developer with the weakest security knowledge unless the appropriate processes (e.g. code reviews) are put in place to ensure more experienced eyes can review developed code. This demonstrates the knowledge gap that exists between the security knowledge a developer needs to have and the security knowledge that the developer actually has. Unfortunately, there are so many different attack scenarios it simply is not realistic to be able to envision and counter them all. Some applications of process can correct this, but in general, the observation holds.

A third possibility is that designers and developers are unable to envision all possible attack scenarios and therefore do not attempt to counter them. Most developers creating products would say that they are creating the securest possible software against the threats that they can envision. In this scenario, the security of the software is tied to the security vision of the developer.

The fourth and final possibility is that the vendors of software accentuate software delivery over software security. This is also somewhat of a likely scenario given the time and resources required to effectively remove the majority of security vulnerabilities that exist within products. In any event, our software development processes must change to prioritize the development of more secure software. We will likely never achieve the goal of attaining completely secure software, but we can improve the overall security level of the software we do develop.

Security can often be breached by simply combining different concepts or technologies that had not been considered by the securing entity. History has shown time and time again that even the brightest people are unable to consider the virtually infinite combinations of threats that can be levied against a given piece of critical infrastructure. One of my favorite examples is from a movie where a retina scan-based, biometric authentication device was used. This might, on the surface, seem to be an invincible authentication strategy, since retina patterns have been deemed, for all intents and purposes, to be unique. What did the criminal do? He literally removed the eyeball of an authenticated individual and used it for short term access to restricted areas. While this seems far fetched in ordinary life, we must remember that we are dealing with criminal elements, many of which are fanatical in nature and willing to give up their own lives to hurt or embarrass the United States and its largest bastion of capitalism – American companies.

Even when a threat is recognized and a plan to deal with it created, often an attacker can simply find a way around it. Consider for a moment encryption. Initially, small keys were used to encrypt data and hackers could break the code by brute force attacks (a brute force attack is simply executed by trying all potential keys until the correct one is found) using high speed computers. Security conscious individuals responded with larger keys and more complex algorithms. What did the hackers do? They began to look at different ways to amass computing power and broke the problem down into parts to deal with the larger keys. Although supercomputers may cost millions of dollars, assembling a large group of personal computers and partitioning the problem onto many smaller computers was a worthwhile effort, was economical, and was rewarded. In fact, there are so many old personal computers out there they can be often had for virtually nothing. Assembling 1,000 node or 10,000 node personal computer clusters is no longer expensive or unrealistic. And since breaking a code is something that a computer can do effectively, computers can be set up to work on these types of problems 24 hours a day 365 days a year. If they get lucky enough to compromise a key, that's great from the perspective of the entities attempting to crack passwords, if not, very little is lost. They have nothing to lose by trying. As we continue to lengthen the keys, the job of breaking them through brute force attacks, takes a very long time. (a quick, back of the napkin calculation, revealed it would take an average of about 540 quadrillion millennia on a machine or machines that can do about 10 billion key checks per second operating on a 128 bit key).

As the keys lengthen, where then do the hackers turn? Hackers then turn to attacking likely keys. What does this mean? It means that rather than having to brute force attack all possible keys, they look to attack keys which are more likely than others – ones which may be based on words, phrases, number sequences, words with substitution algorithms applied, etc. This allows keys to be cracked without the need to try every key. Since very few keys are based on totally random events, this can shortcut the code breaking process. Coming up with a totally random key is hard to do. While there are sources of truly random numbers, computers seeded with a specific value (such as time) are not sources of truly random numbers. At best, depending on the seed value used for the random number generator, you may end up with a PRN (Pseudo-Random Number). There are a number of different seed values that can be captured and used to generate PRN's including system interrupt arrivals, keyboard typing, and others. The key thing to remember here is that encryption is not foolproof. SSL V2 (Secure Sockets Layer Version 2) for example, an encryption technology, was broken in about an hour once the nature of the seed value was known.

The Battlefield

The battlefield on which this war is wagered is both an electronic and physical. Many of the perceived boundaries of the battlefield exist only in an individual's mind. A myriad of different attacks can be launched through the network using information collected from a number of different sources. The majority of attacks are launched through electronic means facilitated by ancillary activities enhancing the chance of success of these attacks. Many of these ancillary activities that occur in cyber warfare also occur in a traditional war. There are surveillance activities, reconnoitering activities, misdirection, false initiatives, diversions, attacks from multiple directions, scouting of enemy resources, looking for weaknesses and so on. These entities spend large amounts of time reconnoitering the Internet looking for weaknesses in a company's defenses. In a recent study, it was estimated that 85% of the attacks were probing attacks, completed as a prelude to a real attack. Those entities with criminal intent can be patient and methodical. Their goals are not necessarily time sensitive.

The main difference is that in this electronic battlefield, the assaults are largely launched from obscure locations and by unknown individuals over distances that many traditional, modern weapons can not reach. Thus, this is not a traditional battlefield and as such, a whole new set of tactics is required to effectively wage war upon the electronic battlefield.

The battlefield on which the war is fought is ever expanding. It is not a stagnant battlefield, but a living, growing, and changing battlefield. Thousands of new users are being added on a daily basis, some with legitimate intentions, and some with criminal intentions. How do we separate out the legitimate from the illegitimate? This is an excellent question and many wrestle with it even to this day. In a traditional war, there are often clear lines of demarcation between territory held by an enemy and territory held by friendly elements. In this war, there are few lines of demarcation. While we may know that there are certain sources of attacks, many times we can not tell whether our enemy is across the globe, across the nation, across the neighborhood, or across the aisle.

Enemy identification plays a major part in any traditional war and yet it is one of the very hardest things to do in an electronic war. Imagine a traditional war where we could not tell the enemy forces from the friendly forces. There are "electronic signatures" or "patterns" that may allow us to conclude that a user is an enemy, but with the ability to move, to effectively change identities quickly, to hide their electronic identity and location, it can become very hard to even find an enemy. Novice hackers tend to make mistakes that allow them to be identified and perhaps located quickly, but more masterful opponents are not so easily entangled.

If I am a criminal element and want to do harm to the denizens of the Internet, the first thing I do is get connected. If I get shut down for malicious activity or my ISP gets shut down because of my malicious activity, I can simply move to another ISP, another state, or another country. While casual hackers may not be willing to go through these extraordinary measurers to get reconnected, professional hackers will do these things. Even if the ISP's get to the point where the information sharing between them becomes effective, I can always use another identity to get connected. In general, there is too little authentication in getting access to the battlefield. If I am truly desperate, I can turn to more exotic measures to obtain access. If necessary, I can even go so far as to get an account in another country, setting up a small presence there and then manipulating those systems remotely so that it looks like the attack came from a source external to the United States when in fact in was launched from here. If I am concerned about getting caught, I can always invoke a program to delete all information on the machine making the attack. Subterfuge and intrigue are rampant on the Internet and there are so many ways to mask things that it makes it very difficult, but not impossible to catch the lion's share of the perpetrators.

The battlefield changes in other ways. Each day, new technologies are introduced to the battlefield, some of which make it more secure, some of which make it more vulnerable. National law enforcement agencies continue to develop new tools and layer on extensive tracking and auditing software to be able to understand what is happening on the Internet, but it is hard to do considering the changes. Some of the ways in which the Internet is extended make it easier to attack. All in all, the Internet is a very dynamic battlefield and there are more changes in store. New protocol changes are on the horizon in the form of IPv6 which provides encryption at the lower layers of the OSI stack. New technologies and operating systems are being developed, new web based protocols such as web services are making their advent and new standards are being developed. If we can say one thing about the battlefield that we fight on, it is that it is in a continual state of flux.

The Network

The Internet is a worldwide system of computer networks that are connected together. These networks literally reach around the globe. The network itself is very compartmentalized; if a part of the network is destroyed, most of the rest of the network can continue to operate. The network itself almost exclusively utilizes a primary network protocol – IP, but utilizes a much larger number of higher level protocols. Countries and different entities connect to the network, sometimes in a way that looks to be perfectly legitimate, but with the sole intention of financial gain or harming other entities. There are a large number of ways to connect to the Internet. Entities can connect through ISP's (Internet Service Providers), ASP's (Application

Service Providers), wireless networks, phone companies, etc. Entities connect through high speed and low speed links.

Some entities have one system attached to the Internet, some have many. Through various technologies, the number of hosts indirectly connected to the network can be obscured until they are needed. The entire network is an unknown quantity. That is, there is no full knowledge of who is using the Internet or what they are using it for. The Internet is open 24 x 7 and there are plenty of opportunities to launch attacks against unsuspecting entities. There are an ever increasing number of entities on the Internet some with benign intentions, some with hostile intentions. Each day that passes sees the Internet expand in a myriad of ways. Each day, new users are added to the Internet, some through higher speed access such as DSL (Digital Subscriber Line) or Cable Modem. Each of these new channels could be used as a launching point against the Internet and its denizens.

The nature of the network itself is both a blessing and a curse. Because of the distributed nature of the network it is extremely resilient to localized failures. If a problem occurs with one part of the network, often the failure can be routed around. Today, even if the local access is disrupted by the cutting or damaging communications lines, a surrogate wireless network can often be quickly established to provide connectivity. In developing countries, many have elected to rely almost exclusively on wireless networks.

On the negative side, the distributed nature of the network allows entities to launch attacks from virtually anywhere, near or far. Where unsecured or poorly secured wireless networks exist, a user in a car can literally drive into the range of the wireless network and connect, obtaining access to the Internet. Wireless networks enable another set of problems including the jamming of the radio frequency spectrum of the wireless network, cracking the WEP (Wired Equivalent Privacy) encryption protocol, intercepting and modifying packets, and inserting new traffic into the network. Once again, this can be insidious because an entity with criminal intent can attack or use the network from a remote location.

If we really wanted to continue to use the Internet as a medium for electronic commerce, a hybrid strategy might prove to be more effective. A series of private, critical infrastructure networks could be developed as the primary exchange conduit for sensitive information with only a couple of different access points to the Internet. At least in doing this, we would limit the number of points from which an attack against the critical infrastructure networks could be launched. The way the Internet is currently designed there are too many places from which to launch an attack, too many access points to effectively control. This is the same concept that is used on a hostile border where entry or exit to a country can be obtained only through checkpoints.

The Weapons

The DARPA (Defense Advanced Research Projects Agency) was the originator of TCP/IP (Transmission Control Protocol / Internet Protocol), a protocol developed to allow different types of systems to communicate. Use of TCP/IP was relegated mainly to military and academic use in the late 1960's, 1970's, and 1980's. In the 1990's however, TCP/IP usage accelerated dramatically to the point where it is the dominant protocol in use today. Whether we like it or not, TCP/IP has become the primary enabler to allow entities to break into systems, but it is not the only enabler, just the door. Standardization when combined with an open network is an invitation to break into systems and networks, unless the standardized technologies are completely secure.

Imagine for a moment what the potential for fraud might be if we simply utilized a standard for credit card numbers. Consider the following credit card numbering standard using a 16 digit number:

Digits	Meaning
1-2	County of Consumer
3-4	State or Locality of Customer
5-6	City or Town
7-10	Issuing Entity ID
11-16	Sequential Number (0-999,999)

If you knew this was the standard, how hard would it be for you to come up with a valid credit card to forge or use? This is the same idea behind TCP/IP and its associated network services. There are standardized protocols with standardized ports, standardized services, standardized addresses, and standardized vulnerabilities. Thus, a group attempting to break into a network has a pretty good idea of what to look for. What is even more insidious though, is that if you believe you have adequately have protected your system from direct, frontal attacks, there always other ways to get to your systems. How? Often, your systems can be penetrated through your business partners or through the systems of others, who may be operating in a trusted environment.

With the acceleration of off shore development, a new set of challenges have emerged. Almost all off shore development organizations require connections to the organizations that they provide development services to. In many cases, these are not private network connections, but rather connections through the Internet. Even if a private network connection is used, the off shore organization likely has access to the Internet. All it takes is a poor security model at the off shore organization to compromise the information and technology of the company that the off shore development organization serves. Because these off shore organizations are not local, there is often no thorough investigation of the security models that off shore organizations use, nor an understanding what commitments to security that an off shore organization has. Since developers often have access to sensitive information, this can introduce significant risk in trying to protect the local company's informational resources.

Vulnerabilities come in all sizes and shapes. Many of the vulnerabilities are those that can be launched through IP, others are launched against technologies that utilize IP. These vulnerabilities can be launched through IP into infrastructure components such as web servers or database servers. More than 50% of the vulnerabilities identified are considered easy to exploit meaning that a hostile element with relatively little experience can breach the existing security. Almost 80% of the total vulnerabilities identified are considered moderate or severe, meaning the vulnerability is significant enough to allow considerable damage. Many vulnerabilities in technologies remain which still have not yet been exposed. Have we seen any signs that we are finally getting to a point where new vulnerabilities have been stemmed?

What are the weapons of the attackers? Our adversaries use a variety of weapons in this cyber war. They use technique-based attacks such as buffer overflow attacks, heap format strings, scans, spoofing, etc. They use scripting and code samples developed or obtained from hacking sites. They use packet sniffers, network probes, and scanning equipment. They use research obtained from legitimate vendor sites, hacking sites, and vulnerability notification sites. They use social engineering skills in order to obtain key information that can allow them to bypass technology protection. Today's Internet attacker has a considerable number of weapons at their disposal and they can use them all. All of this is occurring in a landscape where the vulnerabilities are quickly multiplying. More and more vulnerabilities show up because there are more and more new technologies on the market that were designed without the proper security. Even our existing technologies constantly undergo change exposing new vulnerabilities in what might have previously thought to have been safe.

Our adversaries will continue to expand their use of social engineering attacks. Why? Because successful social engineering attacks can yield information that renders technology protection schemes impotent. Social engineering attacks generally center on getting access to information that they have no business getting access to, usually through subterfuge or intrigue. One of the famous hackers of the past, Kevin Mitnick, testified that he could get virtually all the information he needed through social engineering attacks.

We, if we are not careful, can place a powerful weapon, namely information, in the hands of our enemies. There are so many different social engineering attacks that can be levied against an organization that it is frightening. I have included a short list of common attacks that I put into this book. I have deliberately left out some of the more esoteric attacks since I don't know if they have been tried at this point and don't want to encourage our enemies who may read this book.

The information below, as well as the appropriate countermeasures, is available within many articles that are available on the Internet.

Simple External Social Engineering Attacks
Conducting surveys as someone else
Dumpster diving for specific computer or organizational information
Checking computer media thrown away or sold
Stealing access tokens from company individuals
Calling for a password or ID change posing as someone else
Posing as a law enforcement individual pursuing a cyber criminal
Hacking into a system and requesting sensitive information through E-Mail
Posing as a company employee using collected information
Asking a person to install software to correct a non-existent problem
Typing subjects in E-Mail luring people to open them
Threatening action against an employee for failure to accede to a request
Asking for sensitive information posing as an official or corporate officer
Direct observation (perhaps from afar) of individuals logging into a system

In the past, criminals utilized more mundane types of techniques and items to commit crimes. Those techniques included eavesdropping, forgery, stealing, perjury, etc. Often, simple pen and paper, combined with self-confidence facilitated these crimes. Now, the new cyber crimes committed utilize electronic means to instigate the same type of crimes. Crimes of the past were mostly committed proximate to the victim. The crimes of today can be launched remotely, perhaps thousands of miles away, without ever getting near the victim. Criminals today intercept and alter messages, access accounts electronically, access systems without authorization collecting valuable information in the process, collect information that allows them to steal the identities of others, access medical and criminal records enabling them to extort money from individuals and the list goes on and on.

A couple of years ago, an individual who was caught "trolling for vulnerabilities" had developed a program that searched through servers on the Internet looking for specific vulnerabilities. Once those vulnerabilities were found, he was able to load a program of his choosing that gave him even more accessibility to the systems he had hacked into. It is possible to programmatically seek out weaknesses to exploit. Although our adversaries have the experience to hack systems one by one, this isn't the most effective. Criminal elements prefer more automated types of attacks, at least as a first level attack, since they can cover many systems very quickly. Once they have discovered an initial vulnerability, more invasive hacking activities are undertaken. While this person was caught, most of the evidence against him came from his own computers including logs discussing attacks with other hackers. While many younger hostile elements might make this mistake, more organized hostile elements will likely not.

The tools of our enemies are evolving. Not only can some of our enemies scan for vulnerabilities in a programmatic manner, but they can also launch an attack immediately and programmatically during the scanning process. This means that the window of opportunity between when an exposure is found by an attacker and when it is exploited by an attacker is going to zero. The most recent generation of worms and viruses are self propagating. This means that an individual can launch the attack and then disappear into the background, letting the malicious code do its own damage. Our enemies are also beginning to use distributed attack tools that allow attacks from a large number of systems at once. While this type of attack was launched against the Internet's root servers and the Yahoo web sites, things will likely get worse. We envision the capability of our tools to dynamically respond to changing network traffic. The next generation of hacking tools will likely be of the same nature, being able to try an array of different attacks and dynamically altering the nature of those attacks, masking the very signatures of those attacks, until they succeed.

We share some of the blame for the weapons that they use. In some instances, we forsake good security practices for convenience or for simplicity. Passwords are an example of this. We, as humans, tend to be good at word association. Thus, we often use passwords that are relevant to information about ourselves or are words that are relevant to the business we are operating within. There are different tools that use birth dates, dictionary words, and common names to try to crack passwords. Once access to a network has been gained, it can be only a matter of time before a password is cracked. This is especially true if the underlying system doesn't invalidate a user account after a certain number of incorrect attempts.

The United States still has one of the finest higher education systems in the world. Doubtless that some of the very individuals that we educate will turn against us. Thus, we place another weapon in the hands of our adversaries – knowledge. Once they have the basics, they can, through testing, through development, and through research, develop new and damaging ways to attack our networks and systems. This has occurred in other places in life. Some of the very fanatics which oppose the philosophies of the United States and the western world have been educated in western countries. It is indeed an irony that some of the very people that were educated in the United States have developed some of the strongest dislikes of the United States.

Most of the weapons that we have discussed have been logical weapons (e.g. information). There are other types of weapons that can be used as well. Physical weapons can also be used to compromise facilities, splice into fiber optic cables, intercept radio transmissions, jam transmissions, etc. As the nature of the Internet expands, more and more physical attacks also become available to our adversaries. A well coordinated physical attack on key sections of the backbone infrastructure could render parts of the Internet useless. To this point, most Internet access problems are geographically localized outages that result from accidental cable cuts caused by construction or the like. As we publish and document diagrams depicting the Internet infrastructure, we hand our enemies another weapon: information to attack the physical infrastructure.

The weapons will continue to evolve. What if a company, with a large installed base of software, builds a piece of software in which a developer put a piece of code in the software set to "go off" at 0400 GMT at some date in the future? What might the potential impact be? Given a wide enough install base, the impact could be unbelievably damaging. That might be an ideal time to attack a country. The United States would be hurt considerably if none of its computers functioned at a critical time. These things can be engineered into software. The same could be done with other software makers, but the problem is not just limited to software. What about embedding the same type of "bombs" in the bios of a computer or perhaps modifying chips on the network cards to change or add information to packets leaving the computer. We are talking here about industrial sabotage, but this will also be a weapon in time. The weapons of this cyber war are no different than any other war. Weapons will continue to evolve over time as will the attack strategies. Hybrid attacks (combinations of physical and electronic attacks) will emerge as well and the damage that the weapons can do will continue to increase as well.

Our Enemies

The Internet of course, is far from a secure network. In fact, the Internet is the made up of a very wide variety of entities, individuals, organizations, governments, etc. From this perspective, the Internet is attractive – it can be thought of as a conduit to reach millions of people. But, who really makes up the Internet population? This of course is the $64,000 question. Nobody really knows who makes up the entire Internet community. We know about many of the legitimate interests on the Internet, as they operate in a relatively open manner. It is the unknown, hidden enemies that operate from covert locations, with hostile agendas that we don't know much about.

Consider how long it currently takes to find hostile elements on the Internet even after their attacks are launched. In many cases, the hostile elements are never found, only the remnants of their work testify as to their existence. In addition, except in the rarest of circumstances, hostile elements are not discovered before they do damage.

Many people associate threats to the Internet with individual hackers. In reality, individual hackers do pose a threat to Internet security and the exchange of information, but far larger threats exist. There are many entities which are well organized and well funded entities having the resources to mount serious threats against the Internet and its community. It is these enemies that we need to fear most. On October 21st, 2002, the largest ever distributed denial of service attack occurred on the 13 Internet DNS root servers which provide name resolution. The root DNS servers are the servers at the top of the DNS tree.

Although all 13 root servers were attacked, seven of them were most severely impacted. These included the servers at Verisign (two different servers at the same location), the U.S. Department of Defense, the U.S. Army Research Lab, the Japanese Wide Project location, Autonomica in Sweden, and Network Coordination Center in London. Since it was a coordinated attack against multiple servers, there can be no misconception about the intent of the attack – the perpetrators were attempting to shut down the Internet. The attack lasted only an hour, and although few users were affected, security experts agree that a longer attack of the same nature would have begun to affect users by preventing connection to Internet resources (web sites and hosts) that were referenced by name. Chris Morrow of UUNet, an Internet infrastructure provider, said "This is probably the most concerted attack against the Internet infrastructure that we've seen".

Your enemies are out there, flexing their muscle, expanding their capabilities, developing code in secretive locations that defy pre-emptive identification, reverse engineering security technologies, and learning what does and does not work. It is only a matter of time before they are successful in affecting large portions of the Internet. As of the writing of this book, the perpetrators of the attack on the root servers are still unknown. This attack was impressive since multiple systems coordinated in the attack. 2003 was one of the worst years for worms and viruses and 2004 promises to be even more of a roller coaster ride. By way of cost analysis, a London based security and estimated that combating the SoBig virus resulted in more than 30 billion dollars of productivity and business losses. Whether the amount is actually this high or not, it depicts the costly nature of trying to protect ourselves from an ever evolving enemy.

Enemies on the Internet can attack with relative anonymity from unmonitored locations, from mobile systems, and from protected sanctuaries. They can attack without being seen and often, without being caught. They can disguise their location and the address from which they launch their attacks. They can launch an attack from a location and then walk out the door, never to return, perhaps even taking their "weapon" with them. Although we have been able to catch some of the more immature hackers, professional hackers are not so easily entrapped. Often greed or a personality disorder such as megalomania, leads cyber criminals to be reckless, enabling their capture. We can not count on these flaws to catch those who are making a living off of hacking or those who are operating as a part of state sponsored hacking. If cyber crime is their profession, they have the ability to execute it with surprising efficiency making it tough to catch them.

The nature of the network landscape and the standardized protocols facilitates this. In April of 1999, a person using his home computer accessed the Department of Defense Logistics Agency systems by using "telnet proxy" which enabled him to present his computer as a computer than belonged to the government. Once "in", he proceeded to use these government computers to attack a large number of other governmental computers. He deleted files, changed files, and installed new software which allowed him to capture even more logons and passwords. He was then able to access the systems of Northeastern University and stole large amounts of personal information including social security numbers. One person, with very limited resources, did this in the span of four months. Imagine what a well organized, well funded entity could do.

A study by the Rand Corporation in the mid 1990's concluded that it would be absurdly inexpensive to embark on a cyber war. It takes as little as a single person with a single laptop computer to wreak havoc with a network, even a large network. Although the capability of a single laptop is relatively limited, often due to bandwidth limitations, it can be used to hack into another system which is connected to a high speed network. Once there, a myriad of attacks can be launched through the surrogate system that was hacked into. If the system hacked into is on a trusted network, even more damage can be done.

75

Then other systems on the trusted network can be used to launch attacks. Yes, even a single laptop, though it may be limited by bandwidth, can be used to access systems through which high volume attacks can be launched. As we continue to evolve the capabilities of the laptop, they continue to become more formidable weapons.

When you see a person with a gun, you are able to identify the threat and take action as appropriate. Internet crime is much more insidious. Your enemies are hidden. You have no idea where they are. You have no idea if they operate from one county or from many. Your have no idea if you enemies are collaborating against you or attacking individually. You have no idea as to what their capabilities are or what resources they can marshal against you and your information security strategies. You do not know what their goals are. You do not know when they will attack or what information they have already obtained about you or your company that can be used at a later time. You don't even know if they have someone on the inside, someone that could betray the organization, through the leaking of information, direct sabotage, or worse. Sound like paranoid cloak and dagger stuff? I assure you it is not. And even though some of the hostile elements existing on the Internet have not reached this level of sophistication, all of these techniques will be used in the future. Your enemies are devoting themselves to be able to defeat your security measures. In some locations, they practice their trade many hours per day seeking any vulnerability, any weakness. Consider your efforts to protect your information. Are your resources as driven, as hard

76

working, as persistent as your enemies? The information mega-house called the Internet attracts all comers.

One other thing to consider when facing our enemies is their motivation and determination. Since the payoff for their success is great, they are investing themselves in learning, in testing, in training, and in exploring. Outside of the United States there are dedicated government programs, collaborations of political adversaries, university research, and other similar activities designed to probe and expose weaknesses in security. Now that we have established the nature of some of our enemies, let's look at some of the specific types of entities that oppose us.

Criminal Elements

Criminal elements run rampant throughout the Internet. Their primary goals are illicit financial gain either through direct assault on financial resources or through the laundering of money. In addition, they are interested in information that allows them to protect their existing lines of business. They seek sensitive information such as social security numbers, bank accounts, personal information, credit card numbers, and other information that can be used against people, perhaps even through blackmail. They perform identity theft as a means to an end, getting access to money through a variety of means. Some criminal elements are likely sophisticated enough that they consider the potential ROI (Return On Investment) and the chances of getting caught before engaging in some of their cyber crime activities. Other criminal elements are interested in intelligence information on law enforcement activities and other information that may be related to their ability to operate and extend their core businesses such as pornography, drug distribution, smuggling, stealing, etc. They may also seek personal information on individuals in law enforcement or the court system in order to bend those individuals to their will.

Many of the criminal elements on the Internet are well funded. They have the means to acquire high-speed computers and persuade morally bankrupt, but brilliant individuals to participate in realizing their goals. There are many different criminal elements in existence today. Consider the Columbian cartels, the Russian Solntsevo group, the Sicilian mafia, and the Palestinian PFLP-GC group. This is by no means an exhaustive list of these types of elements, but belies the nature and diversity of the criminal elements roaming the Internet. The criminal activities practiced by these groups are lucrative in nature and enable them to expand their criminal activities to include cyber crime.

Just as businesses turned to computers in the 1960's to effectively manage their business processes, now criminal groups are doing the same thing. They use their acquired technology to perform additional criminal activities. Criminal organizations, while they exist in the United States, are even more rampant overseas where there are often no cyber crime laws. The Computer Crime Research Center, in a study conducted in 2001, estimated that as of the year 2000, there were 960 well organized criminal groups within the Ukraine whereas there had only been 265 recognized in 1991. It is envisioned that by the year 2010, there may actually be criminal states – governments providing safe havens for criminal elements and supporting cyber criminal elements through direct funding and immunity from prosecution.

Far too often we underestimate our adversaries' abilities in this area. We assume that America's technical prowess is sufficient to defend its information resources against any attack from less educated countries or individuals. We are generally a complacent society when it comes to security. We have lived most of our lives in relatively security. Only when the three planes slammed into three different buildings did we begin (and I stress begin, since there are still so many vulnerabilities in the air transportation system) to take airline security seriously. This was a wakeup call to all of America that we can no longer take security for granted.

Employees

An often overlooked source of threats, are employees and former employees. Employees and former employees may have access to a wide variety of sensitive information and they can use that information to further their own goals or can simply pass information to others who may use it. In addition, they are resources that can be pressured. Employees that are pressured through some means may be forced to reveal information that pertains to their environment which may lead to a compromise of a company's environment. Although we have seen few landmark cases of an employee being pressured to reveal information, this will eventually occur.

The bigger source of problems is disgruntled employees – past or present. They can use their job related information to launch attacks from inside a corporation that can do more damage than an external attack. Often, they have an additional advantage in that they can circumvent existing auditing and logging procedures erasing any trace of the crime, especially if they also have access to logging facilities as well. This can make it hard to prove the existence of such an attack and can make it hard to prevent these types of attacks in the future. While employees must have the tools and authority to do their work, there are ways to deal with employee security breaches. Regardless of whether an employee is pressured or disgruntled, they can be a source of critical information that can be used to circumvent security models and technologies.

In addition to this, employees are the number one source of errors. Errors are not introduced intentionally, but they can be as damaging as any of the more overt external attacks. All it takes is an errant configuration in a piece of software or hardware and data may no longer be encrypted, an administrative interface may be left active, traffic is routed to a location that it has no place going, or ports may be left open on the corporate firewall. This is a hard one to deal with since no one has perfect employees. The only approaches to dealing with issues like these are layered security models or moving electronic commerce to a more forgiving landscape, providing some insulation from these types of failures.

Employees are always a prime consideration when developing security protocols, but most of the issues that can arise from employees are dealt with through social engineering and not purely technology. Technology can however, be used to keep audit trails of activities within an organization.

Employees that develop software and scripts (e.g. Systems Administrators) also have the ability to modify their code to perform malicious activities ranging from Salami Slicing (taking extremely small amounts of cash – the infamous always round down issue) to embedding Trojan Horses in software. Once Trojan Horses have been embedded in software, they can perform malicious activities such as capturing and sending out passwords, capturing account numbers, etc. This type of criminal activity can be very hard to catch in an organization unless the proper social engineering protocols are in place. In some shops, this type of activity is thwarted by routine code reviews convened by multiple members of a development team, insuring that embedding of malicious code can occur only if the entire team is corrupted.

Fanatics and Terrorists

Fanatics also roam the Internet and not just one type of fanatic. There are many different types of fanatics. Religious fanatics, performing horrible acts in the name of God, military extremists fearing their governments, and rebellious insurgents bent on toppling governments, all exist on the Internet. They seek information that can be used against their adversaries. In many cases, they are willing to sacrifice themselves to hurt others. Laws don't apply to them in the traditional manner. Our system of law is based on the basis of deterrence. If the punishment for a crime is bad enough to make you think twice, perhaps it will stop you from committing the crime. This works for most crimes with the penalties being either financial or freedom oriented or both. Fanatics do not consider this deterrence. They are more interested in the end goal regardless of the consequences.

This has some profound implications for social engineering. When the law and its consequences are no longer a deterrent, a whole new set of social engineering problems arise. Terrorists could break into a company to get a hold of sensitive information, making it look like a routine robbery and then abducting an employee, holding them until they extract sensitive information from that person. We must be on guard about letting our system of values bound what potential adversaries may be willing to do to us or our resources. It was very likely the set of values of the passengers on the planes which slammed into the various buildings that prevented them from considering the eventual outcome.

As evil deeds in the name of religion continue to make their way to center stage, we should think about the unthinkable. The very things that we take as givens will be put to the test. Consider this question. If your lead security individual stopped coming to work, how long would it take before you revoked all their authority and removed all of their access to systems and networks?

Hackers and Crackers

Hackers and crackers often don't have a real goal in mind; instead, they view every web site as an invitation, every security construct as a challenge, and every application as something to be broken into. They often wander aimlessly from site to site, seeing what they can and can't get into. Often hackers and crackers get into trouble because of what happens after they are able to break into a system. Sometimes the damage they cause is only mischievous in nature such as changing the contents of a web site to reflect a political opinion or a sports related opinion. Unfortunately, there are times that when the opportunity for financial gain presents itself, a hacker or cracker can not resist the temptation and they turn to a life of crime. Many hackers and crackers also suffer from a form of megalomania, seeking to prove their abilities superior to those they attack.

Hackers and crackers, once they have become proficient, have a couple of different paths to take depending on their morals. Some turn to ethical hacking and making a good living in the process. In this role, they try to compromise systems without doing damage, reporting on the failures of organizations to secure their resources. Some hackers turn to a life of crime once they have enhanced their skills. These individuals may be hired by criminal groups or they may act independently, but either way, they have decided to walk the dark path, using their skills for harm rather than good. Hackers and crackers often are devoid of personal relationships spending much of their time on the Internet learning and attacking and learning and attacking. In many cases, hackers and crackers can be socially inept. Unfortunately, this is not a foolproof way to identify them.

Hackers and crackers, even if they do not utilize the information obtained through their exploits, often have the opportunity to sell the information to others who will use it for criminal activities. In addition, many actively read and contribute to hacker sites on the Internet. In these sites, code samples, known vulnerabilities, and sometimes company specific information is kept. They constantly monitor vulnerability tracking sites, hoping to obtain critical information that they can exploit before companies can close the window of vulnerability.

Many hacker groups are well organized into secretive groups that share information about vulnerabilities and attacks on a regular basis.

Examples of these groups include globalHell, cDc, Hactivismo, G-Force, #conflict, total-kaOs, the Darkside Hackers, the Pakistani Hackerz Club, the Chaos Computer Club, and the Legion of Doom. They develop detailed software that exposes vulnerabilities in software all the while protecting their own web sites. Some are political advocates, with stated positions against net censorship and attacking entities whose opinions are different than theirs. When individuals of this type combine into groups, they can become formidable adversaries. Knowledge is often blended together across multiple members of a hacking group so that attacks which might not have been easily orchestrated by a single individual can now be implemented using the knowledge from multiple individuals.

What might be most scary about many of these hacker groups is that they tend to be made up of young and idealistic individuals whose conscience does not impair them from hurting others. Many of these individuals live their life in computers, spending up to 20 hours per day on the computer.

Hostile Governments

The most recent entity using the Internet to attack information resources are hostile governments. In some cases, the government participates and is actively involved in malicious activity, in some cases they sponsor it, and in some cases, they simply turn a blind eye to the activity within their borders. Governments involved in this type of activity have considerable resources and can use these resources to research cyber warfare on a global scale. Jung Beom Seo, CEO of Defense Korea affirms, "The cyber terrors we have seen are nothing when compared to the incoming war." What experts worry about is a future cyber war which can be conducted by individuals against a common enemy, as well as against the military of a specific country.

Each nation must prepare for the future war including defending against the spreading viruses or disturbing systems over the Internet and military networks. Kwang Hyung Lee of the Korea Advanced Institute of Science and Technology (KAIST) advocated that Korea needed to train 500 hackers in 1997. Later he realized the requirements were growing so quickly, that the government needed to train ten times more professionals than originally thought.

Countries have as their primary goal, accessing and using information for a variety of motives. They seek information about national security, resources of other countries, critical infrastructure locations, war plans, troop deployments, information about how to build weapons, how to defend against weapons, etc. With information of this nature obtained, they are better prepared to attack critical infrastructures attempting to wound their opponents politically, economically, or militarily. Countries sponsoring terrorism currently on the Cyber-Terrorist watch list include Cuba, Libya, Iran, Iraq, North Korea, Sudan, and Syria. Countries that have active terrorist activities within their borders, although not necessarily sponsored by the government include Lebanon, United Arab Emirates, Saudi Arabia, Pakistan, Kuwait, Afghanistan, Egypt, Indonesia, and Jordan. Some of the countries above have considerable resources from their oil revenues, others are evolving. We have seen in the last year a dramatic increase in the number of attacks originating from these countries, some as much as threefold over the previous year.

This author can easily envision a dramatic increase in these types of attacks. Some governments have already developed entire organizations dedicated to the premise of cyber warfare. Cyber warfare is more common than traditional wars, although we have yet to see some of the larger attacks. Cyber warfare will be executed between countries even when the constituent governments are not involved in a traditional war. When traditional war does occur, cyber warfare will certainly accompany it. One of the most effective ways to win a traditional war is to knock out the command and control centers of an army. This causes confusion among the members of the militia and leads to inconsistent execution of battle plans. This is precisely what the United States did in the Gulf War, attacking many of Iraq's command and control centers before moving into the country. Resistance is reduced when the command and control centers are compromised or eliminated since it becomes logistically hard to move around troops and equipment in a coordinated manner and to respond to dynamic threats on the battlefield.

Enter cyber warfare – one of the means to effectively take out command and control centers without the need to drop bombs. The more that a country relies on its computers, the more vulnerable they will be to cyber warfare. During an April 24, 2002, hearing on homeland security before the Senate Subcommittee on Science, Technology and Space, Rep. Sherwood L. Boehlert (R-NY) shared his thoughts on cyber security claiming that "Our adversaries are going to get more and more skilled, and we must get smarter and smarter to counter them," he stated. "Like the Cold War, the war against terrorism must be won in the laboratory as much as on the battlefield."

All governments have to invest in this type of warfare. They can not afford not to. Governments, hostile and friendly, need to be able to launch cyber attacks and to respond to cyber attacks from other hostile entities. Information is now king; we have evolved into an information-based society. Information controls our finances, our medicine, our politics, and yes, our wars. Cyber wars can be launched by governments, against other governments, without actually declaring war. When a traditional war occurs, there is generally evidence testifying to that fact that a war is in progress. Either troops are invading a country, or bombs are dropping, or tanks are on the move, etc. Cyber wars can be launched from anonymity so we can't even be sure who is attacking. For example, if an attack is launched from a hostile country and the attack is successful, we may be able to trace the attack to that hostile county, but we may never be able to determine which person or group within that country launched the attack. It may have been government sponsored or it may not, but since we don't have the necessary visibility into those areas of the world, we may never find out. While we do have the ability to selectively shut down traffic from a country, we must remember that since we can not distinguish good from bad, shutting down all traffic from a country impacts legitimate business interests as well as illegitimate activities. In addition, the distributed nature of the Internet may allow them to route around a shut down ingress so that the attack proceeds through other countries.

We are especially at risk from countries such as North Korea which flaunt their nose at the international community, defying mandates as they see fit. It is these countries that no extradition treaties will ever reach and no prosecution of cyber criminals will occur. Just as certain countries still sell arms to countries like North Korea today, so will countries in the future sell Internet access to countries like North Korea. Once the access is available, they can launch attacks from that access. Even if we are totally successful in limiting access from rogue countries, these countries can always invest in their trade in their own country and then go somewhere else and launch the attack.

In a traditional war, it takes tremendous resources and time to reposition military hardware to launch an attack against a country. Even the military buildup, as a prelude to invading Iraq took many months. With cyber warfare it is different. Cyber warriors can work within their own countries, on identical technologies, to develop a devastating worm and then write it to a CD (Compact Disc) taking it across the border where the attack can then be launched. Or it can be copied to remote systems over the network. Remember that damaging code can be moved from location to location with impunity. The attack now appears to come from a different country than the actual country from which it was conceived as an attack. This demonstrates the nimbleness of the cyber warrior. While the development of the worm takes time, it is not apparent to us and so we have no way to prevent it. It can be developed and tested in a small room with no visibility from external scrutiny. When we move large amounts of military hardware around, it is visible to a number of different countries. When cyber warriors (or info warriors as they have been termed) attack, there is no evidence of the attack until they launch it. This type of war requires a whole new type of logistics management.

Even when governments are not hostile to us, we have trouble in prosecuting the perpetrators of cyber crimes. The suspected inventor of the "Love Letter" virus was arrested in the Philippines in June of 2000, but his charges were dismissed when Philippine authorities determined that existing laws did not apply to cyber crime. Since then, the government has enacted a law defining hacking and virus propagation as criminal activities.

Even when criminals are caught by governments and prosecuted, the penalties are often light involving probation, home incarceration, small fines, and in some cases, limited jail time. To this point, we really haven't taken cyber crime seriously enough and in many cases, neither does the rest of the world. Until we get to a point where we consistently deter this type of activity through significant penalties and aggressive prosecution on a global scale, we will continue to be at risk from these types of countries and individuals. It will likely take a significant attack that affects large numbers of entities to stiffen the penalties for cyber crimes. The penalties for grand theft auto can be higher than the similarly valued cyber crimes. An example of this was an individual who was sentenced to six months in prison, followed by a two-year period of supervised probation. He was also ordered to pay more than $10,000 in restitution. What had he done? He hacked into his former employer's servers and deleted files, modified access privileges, sent out electronic mail as an authorized representative of the firm, and modified billing records. Considering the actual and potential damage, this level punishment was rather light.

Our Allies

Although things in law enforcement circles have been improving, pursuit of cyber criminals can be very difficult. There are simply too many places from which to launch attacks, where no treaties or extradition agreements exist and where no cyber crime laws have been enacted. Attacks can be launched from countries with which the United States has a hostile relationship - North Korea, Iran, and Libya to name a few. Even attacks launched from countries friendly to the United States may not be able to be prosecuted. The whole nature of the Internet access model allows individuals, with minimal cost, to obtain full electronic access to the Internet. To this, day, most attacks originate from the United States, but others countries are beginning to close the gap. Although there are a number of allies that will help us in the evolving war, they may not be from the sources that you expect. For the most part the security of your information is largely in your hands.

Even if the different law enforcement agencies had all of the necessary tools and training, they are generally not equipped to attract and retain the best and brightest of the cyber warriors. The security staffs they maintain are often under-trained and over-subscribed. Often, governmental and law enforcement agencies have limited salaries that can be paid to Information Technology professionals and these salaries often do not compare favorably with what a talented security professional can earn in the private sector. Despite the newly founded focus on security since the September 11th attacks, a recent Computer Security Report Card released in November of 2002 concluded that every major agency in the Federal Government exhibits significant security weaknesses. Even more worrisome is the fact that many of the agencies who handled the most sensitive information, fared worse in the most recent review than in the previous review. The Department of State went from a D+ in 2001 to an F in 2002 and the Nuclear Regulatory Commission went from a C in 2001 to an F in 2002. Others fared similarly, showing no improvement. The Department of Defense and the Department of Energy received F's in both years. Relying on the government for protection from the hostile elements roaming the Internet is a dicey proposition.

Governments as a whole have an entirely different set of problems, especially at the state and local levels. Governments are made up of a series of elected, appointed, and hired individuals. In some governmental entities, elected officials are bound by budgetary processes that require support of others throughout the governmental entity. For example, an elected official may require support from a formal budget committee to operate their department or invest in new technology. There are also some agencies considered "cash funded" agencies. Cash funded agencies generate their own cash and often have statutory fees paid to them, minimizing their reliance on others to extend their systems. Elected officials are charged with performing their statutory duties to the best of their abilities and there is wide latitude given to elected officials in how they see best to execute their statutory responsibilities. This can have serious consequences to organization wide security protocols.

Later on in this book, we shall talk about the need to implement security as a series of coordinated activities that transcend the entire organization. If an elected official finds the security protocols of a governmental organization cumbersome to their statutory duties, they may elect to bypass or modify carefully thought out security protocols. This can have dire consequences in a networked world because it takes only one such violation to put the entire governmental organization at risk. The same thing can happen in a private organization as well so the problem is not necessarily unique to government. This is not to say that all elected officials pose a risk to securing governmental resources, but there are elected officials that will operate outside the bounds of best security practices in the name of their duties. Over time, elected officials may have changes made to their swearing in ceremonies where they will have to commit to uphold the security practices of the organization and to protect all information entrusted to them. New statutes will eventually begin to appear holding elected officials liable for protecting the information they act as stewards over. Many states already have laws in place that protect the privacy of information of individuals. Those who do not practice diligence in protecting this information may be held liable from a legal perspective.

The same types of concern exist in the private sector as well. Rogue business units, which operate outside the bounds of enterprise security practices, can compromise corporate security by undertaking actions which circumvent established security technologies, policies, and procedures. While most business units do not actively seek to undermine the corporate security model, the actions they undertake can have unintended consequences.

Process

Process is an important ally in the war. Process can help us defend against many of the social engineering attacks that are launched against our entities. Process is effective when there are few or no exceptions to the process and when the process is very carefully thought out and methodically executed. The problem with most process is that we get into a position where we think we know better than the process and we choose to ignore some or all of it. Consider the documented steps describing how to put a typical plastic model together. How many of use have ignored the formal directions, because we thought we knew better, only to have a leftover part? Developing a well thought out process can be difficult just as it can be difficult to consider every threat against a resource. Nevertheless, there are attacks that can be thwarted by process only and not technology.

There are a wide variety of attacks that can be leveraged without the direct use of technology, but with social skills, and these can be the most damaging since information harvested through these attacks can render technology protection useless. Imagine an exchange something like this, which has occurred in the past. It starts off with the supposed CEO (Chief Executive Officer) calling the help desk to ask for his password to be reset. Note that the person launching the social engineering attack has some knowledge of the organization, perhaps gained through another social engineering activity or something as basic as the company web site.

Imposter CEO: I seem to have forgotten my password to the Enterprise Resource Planning System

Help Desk: Okay sir, could I please have your social security number and your employee number?

Imposter CEO: I have no time to waste on silly questions. Please reset my password to 'ARNOLD23'.

Help Desk: I'm sorry sir. I can not do this without your social security number and your employee number as verification of who you are.

Imposter CEO: Do I have to speak with your supervisor about this? This is a shame that you don't even know your own CEO's voice on the phone. Did you attend our annual Christmas party last year?

Help Desk: Yes sir, but what does that have to do with…

Imposter CEO: Then don't you recognize my voice?

Help Desk: I am sorry sir. The process that your Chief Security Officer approved will not allow me to change your password without the information that I have requested.

Imposter CEO: Very well, I will get someone else to change it.

This is just one example of a very, very simple social engineering attack deflected by process. We might differ on the best information used to validate the individual, but in general, authentication of who we are dealing with, is key in defeating one class of social engineering attacks. It is important to select information that is not generally available to anyone besides the party being dealt with. For process to be successful though, one of two conditions must exist. The first is that all are bound by the process, exceptions don't exist. The second is slightly less dogmatic whereas certain exceptions are allowed, but these exceptions must be independently validated. This case has the person requesting the service (in this case the imposter CEO) providing additional information that must be independently validated through a trusted and known source. Even then, we must be very careful to ensure that the information passed to us by the remote party could not have been obtained through any means. If we have any doubt, the request is not honored. Thus, we can not assume, under any circumstances, that the caller is providing accurate information unless we can verify it independently of what the caller tells us. Here's a similar exchange with the CEO supposedly at remote office, but without the request for personal authentication information.

Imposter CEO: I seem to have forgotten my password to the Enterprise Resource Planning System

Help Desk: Okay sir. I've noticed that you aren't calling from an internal number to this campus. May I ask where you are sir?

Imposter CEO: I am at our Detroit field office.

Help Desk: I see. May I have the number you are at sir?

Imposter CEO: Yes, although the phone is busy here. Please call me on my cell phone at 555-3433 and let me know what the new password is.

Help Desk: Thank you sir. I will call you back with the new password in a few moments.

Imposter CEO: Thank you.

At this point, the Help Desk person calls the Detroit office, the CEO's administrator, or any other corporate officers to validate the CEO's location. If his location is validated as somewhere besides Detroit, the process to change the password takes an exception route in the process so that an alternate set of steps can be invoked. Even if the location of the CEO is validated as Detroit, the process doesn't end. Once the location has been validated as Detroit, the Help Desk person will then call someone in Detroit to locate the CEO and validate his request.

If the information can not be validated, exception path activities may be undertaken. Examples of exception path activities that might be taken include tapping the line in the event that the "CEO" calls back or locking the CEO's account out since this has been targeted.

As we move forward in time, and more protection constructs can be relegated to technology, we can expect the nature and the gravity of social engineering attacks to increase. There is a whole cadre of social engineering activities, some straight out of science fiction movies that will appear in the future. We must be especially on guard against these types of attacks and expect the unexpected. Also, technology will begin to become prominent in social engineering attacks. Technology will be used to augment social engineering attacks so that the likelihood of them succeeding improves. Remember that just because a law may deter us from taking an action, there are individuals in this world which no law will deter. Eventually, there will be devastating social engineering attacks that will reap large rewards.

Process can also help us in protecting us from attacks that are not socially originated. A formal testing process, formal change process, and a formal technology evaluation process are all examples of processes that help us control our environment. These types of processes help us with our technology deployment and changes to existing technology. For example, if a change control mechanism requires, as a part of the process, a formal signoff to approve the change, a back out plan in the event that the change fails, and signoff of the change submitter ensuring that the change has been tested from a security standpoint, we improve our chances that we don't open up any new holes in our computing infrastructure. The change control process needs to include all key individuals in an organization or organizations (for electronic commerce partners) so that potentially impacted entities can be aware of the change. Note this doesn't guarantee that we will be safe, but applying process can help to protect us in a number of different ways.

Time

Believe it or not, time can be a very valuable ally to a company. As we discussed earlier, if information that has a relatively short half-life can be protected through its valuable life, getting access to the information after its useful life will likely have little or no effect. We have gotten to a point that with reasonable due diligence in selecting seed values, it is no longer possible to break keys through brute force attacks unless the hacker is simply lucky and hits upon the key early in the attack (and the chances of this are virtually nil with current technology and keys 128 bits and longer, unless they use a seed value that is easily testable). If entities choose to use simple keys, either intentionally or unintentionally, they risk their data being compromised. There is of course, no guarantee that a way to decode a key through either a special chip (some sort of ASIC for example) or mathematics will not be found in the future.

Time also is an ally in other ways. If the security we implement is not perfect, but is reasonably sound, it can allow us to be able to detect attacks and allow us to respond before resources are compromised. Thus, once the surveillance activities have begun, the clock has started. As a recent report indicated, up to 85% of the "attacks" that occur on a companies resources are launched to perform reconnaissance as the basis for an actual attack. This probing, if it can be detected in advance, may give a company lead time to address the potential of pending attacks. In addition, other governmental entities can be engaged to help look for the entities probing company resources. When a hostile element conducts vulnerability probing activities, there is no way to know how soon vulnerabilities will be found or if they will ever be found, but there is risk to your organization's resources if the organization doesn't take action when the probing is first detected. In order to use time as an advantage, we must be diligent enough to respond to probing and anomalous traffic patterns against our resources before the real attacks are successful. We must spend the time to continuously review our network traffic logs and our system logs to catch reconnaissance activities before they mature into full blown attacks.

Time is also on our side from another perspective. At this particular point in time, many of the adversaries outside the United States are less well equipped than we are, although this point may be contentious with some. Even with this advantage though, some

101

attacks are successful. Over time, their capabilities will continue to increase, thus we still have some time to evolve our defenses in a manner that will provide greater protection than we currently have. There are plenty of financial resources available to those who would compromise our informational resources. We must make sure that we do not squander any advantage we may have at this point in time and must use the time to get our houses in order with respect to security. We can not count on this advantage forever though; now is the time to act.

Time can also be a significant weakness for us if we continue to operate reactively. A traditional war is wagered over days, weeks, months, or even years. Unfortunately, electronic wars can move magnitudes faster than do traditional wars. With the recent advent of current generation worms and viruses, the entire battlefield can be compromised in a matter of hours. When attacks move this fast, there is barely time to publish the vulnerability and react to it. As newer and more nimble versions of worms and viruses attack, we will have likely even less time. All it takes is a Security Engineer asleep at the switch and the organization's resources may be compromised. The time for viruses to spread is heading to 0, and while it will never reach that, the closer it gets to that, the only real remedy that may be effective is to shut down an organization's Internet connection for a considerable period of time. Since this disrupts electronic commerce, we are at a catch-22.

Anticipation

Although we have already seen that it is very difficult to envision and respond to many of the threats against our resources, this is exactly what is required. Consider for a moment the companies providing anti-virus software. Their work is of value, but they are a reactive force in the marketplace. When a virus first appears, it takes a company providing an anti-virus product some time to research and respond to the virus. Since new viruses tend to propagate through the Internet faster and faster, the answer is not to try to respond to new viruses, but to prevent them in the first place. A recently introduced Trojan horse style worm has appeared and it is has been introduced in a manner that makes it appear legitimate. It is presented as the fix for a newly discovered, security vulnerability. For the average rank and file computer user, this can be a nightmare. Instead of securing their systems, they are actually opening up a new vulnerability. If the fix is presented as coming from an authoritative source, how will we know the difference without careful research? The subversive elements on the Internet are now beginning to use our own constructs against us.

With some systems, their very nature precludes looking for a tell tale sign of an infection. It is far better to try to understand the potential vulnerabilities in software, through proactive (i.e. before release) testing. Once different types of vulnerabilities are identified, specific countermeasures can be developed to reduce or negate an attack. In the long term, it simply will not be sufficient to operate in a responsive mode; a proactive mode is required. As of this writing, there are more than 65,000 different virus signatures that exist and the list is growing.

Although computing power is growing according to Moore's Law (computing power doubles approximately every 18 months), the amount of CPU (Central Processing Unit) resources to check for virus signatures on a system is also growing. While there will likely be plenty of CPU power to handle this in the future, it is clear that the files supporting the virus definitions are only going to get bigger and that the response time to check items for viruses will lengthen if we proceed with the current model. Have you ever noticed how long it can take, even on some of the new computers to open a simple document? Why does it take a while? It is because the document being opened is being checked against the virus signature database to ensure the document contains no harmful code.

In addition to this, computer viruses can mutate just like real viruses. Mutated viruses may not exactly match established virus definitions and so they may not be caught by an anti-virus program at a given point in time. Once again, trying to keep up with all of these types of different threats requires continual updates to our virus signatures and we still have the window of vulnerability that exists between the time a virus begins propagating through the Internet and the time that a fix is available and deployed. We need to get to a point where we leverage technologies which can proactively anticipate attacks rather than reactively match them to constantly out of date databases.

Another way to anticipate attacks is through testing. Testing of software can occur through a wide variety of methods and by a diverse set of individuals. If an entity does not have the resources to constantly be testing software, researching vulnerabilities, etc., a third party organization can be engaged to perform the testing. If it is not possible to eradicate, or at the very least greatly reduce the vulnerabilities within a piece of software, it might be necessary to consider replacing the software. Many organizations are unwilling to take this step for fear of losing their investment in the software. This can be a hard decision to make, especially for a product company. The key here is to anticipate, through research and testing, how a particular piece of software might be compromised and to eliminate those avenues of vulnerability before deployment. Software testing is only one component of a security strategy though. It is imperative to be proactive with so many more elements of a security model. It is necessary to establish the necessary social engineering practices, the appropriate change control procedures, and the appropriate research and analysis methods.

Thus, another of our axioms comes into focus. **A proactive approach to security is required to have any chance to thwart many of the threats levied against information resources.** A reactive approach to security only leads to compromised systems and sensitive information being released since it is impossible to protect resource by waging a purely defensive war in a reactive mode. There are a number of different security frameworks on the market today, and while none of them are foolproof, at least they offer a regimen for beginning to define the critical assets and information within an organization and then develop plans to protect them.

The Reinforcements

There are a number of different governmental and academic organizations that work to monitor, manage, and investigate Internet crime. These include the GSA who has been working to coordinate the setting up of response teams to handle incidents quickly and completely, Carnegie Mellon University who operates the CERT (Computer Emergency Response Team) in more than 140 countries, the 3WC (World Wide Web Consortium), Interpol, the Government/Private Sector based FBI (Federal Bureau of Investigation) Infragard Program, the United States Secret Service electronic crimes task forces, FIRST (Forum of Incident and Response Security Teams), ECTF (Electronic Crimes Task Force) and others. Most of these entities are reactive in nature, responding to events that have already happened. Even the proactive work done through the DoD RDT&E (Department of Defense Research, Development, Test, and Evaluation) unit has a fairly small budget to actively develop computer information warfare strategies. According to the Military Information Technology Online, the 2003 budget was even smaller.

The recent publishing of the National Strategy to Secure Cyberspace correctly identifies that "By 2003, our economy and national security became fully dependent upon information technology and the information infrastructure. A network of networks directly supports the operation of all sectors of our economy…". While this observation is accurate, the conclusion to try to secure cyberspace, at least the way we currently define and use cyberspace, is flawed. Many individuals continue to feed the frenzy surrounding the Internet operating from the assumption that we will eventually get to a "safe" or "secure" Internet.

There are so many changes going on in the technology world that there will never be a static technology base to secure. As soon as we figure out how to secure a given technology, a new technology appears with new features, new vulnerabilities, and with new challenges in securing it. Technology is not static. It is not logical to assume that at some point in time we will achieve a secure Internet since the very things that make it up are in flux. While we can be diligent, while we can invest large numbers of resources in attempting to secure the Internet that will never be totally secure, the question is: is it worth it?

Also joining the war, more and more local law enforcement agencies are developing cyber crime investigation units with expanding capabilities. Unfortunately, many of the organizations that are charged with the oversight of the Internet have neither the resources nor the authority to police activity on a global level and can even be handicapped within our own borders. Even if some of these entities have world wide reach, there are simply too many countries that do not share extradition or prosecution treaties with the United States. Nor do they share the same perspective on liability or accountability as the United States. Less than 50% of the countries worldwide currently have electronic commerce statues in place.

The reality is that there is simply no credible oversight organization that validates that once an entity connects to the Internet, it will use the Internet for valid business purposes. In addition, the problem extends beyond this. While many of the major providers of Internet services throughout the world are perfectly legitimate, it does not necessarily follow that all of their customers are legitimate business entities. It is possible to get multiple internet accounts with the most basic of information, most of which is never even validated. Thus, while a person using the network appears to be a legitimate entity, they may in fact be a terrorist or an entity that is accessing the series of networks only to wreak havoc on its other denizens.

Some of the law enforcement agencies are beginning to develop new tools, such as the FBI's Carnivore product designed to inspect and analyze IP packets as they pass through the network. This type of tool is an example of a first generation proactive monitoring tool designed to catch malicious activity in progress rather than respond to the artifacts of an attack. These types of tools need to evolve quickly and to be extended to larger sets of networks. Other governmental agencies are beginning to position network monitors on some of the Internet backbone networks. While these monitors can be useful in deflecting traditional attacks, I fear that they may be insufficient to detect the next generation of attacks. In the upcoming discussion of technology as an ally, we depict several of the flaws in relying only on technology. There are often ways to obscure attacks in the form of legitimate traffic.

Cyber crime laws are also beginning to evolve, but we simply must accept that at least for the time being, some entities will always find a way to keep out of reach of the long arm of the law. In places where cyber crime laws are weak or non-existent, the growth of entities bent on cyber crime is skyrocketing. The Ukraine is just one such example of a place. Even when laws do exist, there are not always clear demarcations of jurisdiction leading to complexity in investigating and prosecuting cyber criminals. Finally, there are some municipalities that will probably never be interested in prosecuting cyber criminals. Although countries pursuing this approach could be ostracized from the global community, this may not be an issue for some countries. North Korea is an example of a country that has few ties to others throughout the world and so may not be affected by the pressure of the global community.

Another source of support in the cyber war is the myriad of different security companies that exist in the marketplace. In general, some of these security companies suffer from some of the same problems that other companies do. Many security companies do not invest heavily enough in training and may not be current enough with their knowledge of what is happening in the industry. Continuous training is required for any security company that is serious about trying to help their customers. Although it may conflict with projected revenue goals, I would hope to see security companies investing an absolute minimum of 10% of an employee's time in training and would feel more comfortable if this number approached 20%. Seem high? Consider the consequences of not spending enough time keeping up with changes in the security landscape. In addition, all training does not have to be classroom style training, but the need to keep as current as possible is important.

Some Internet consulting companies think more highly of themselves than they ought, promising to secure a company's information systems, but without the breadth or depth of knowledge needed to accomplish the tasks. This brings us to another of our security axioms. **Security companies are to be engaged with a clear goal of improving security around informational resources through current information, but not trying to guarantee it.** We should not assume that security companies will end up fully securing our information resources, because they simply don't have that ability. Although they make their living off security consulting and may have more information than internal staffs, they have some of the same limitations in trying to envision and counter every threat as do internal staffs. Even if we engage them at a point in time, the best they can do is to try to secure informational resources at that point in time. As soon as we start making changes, all bets are off.

It is important to very carefully select a security consulting company in today's world. Many of the companies in business today have sprung up over night and have little real world experience in securing corporate information resources, especially across the wide array of technologies that most organizations use. In order to effectively be able to help a company in securing their information resources, a security company needs to have breadth and depth of experience. This means not only technical knowledge, but also knowledge of social engineering attacks, process, and other, non-technical disciplines. Working with an organization to identify and develop process to defeat social engineering activities can be very time consuming for many members of a company. In addition, security companies will often operate with the existing technology base as a given even through it makes their job of trying to secure the environment nearly impossible. One thing that we can do as corporate officers is to candidly ask our security consulting companies for a written appraisal of the technologies that we are using.

Technology, a Fickle Ally

Technology when properly applied can be an ally. Unfortunately it is a fickle ally due to our reliance on it. Technology is flawed. The reasoning goes something like this: humans build technology, humans are flawed, and thus, technology is flawed. In reality, we as humans tend to rely on technology too heavily, often without an in depth understanding of what we are relying on.

Technology facilitates complacency. We build a new technology and we care for it, test it, adapt it, implement it, and then leave it alone assuming that it can do its job. How many times have we seen a failure in technology cause problems? Is it the fault of technology? No, rather it is our negligence once the technology is implemented that causes the problem. Technology in and of itself, is not normally dynamic. When the landscape that the technology operates in changes, the technology also needs to change, yet we continue to assume that it will work. Technology is truly excellent and can serve as an enabler, helping us reach our goals, but it is not sufficient all by itself. In the security war, the entire battlefield is dynamic. There are new viruses, worms, vulnerabilities, attacks, etc. So as the landscape changes, so must the technology. We have learned this with viruses, albeit in a reactive mode. We now know that we can not simply leave a single set of virus definitions in place on a machine; we must constantly update them. We must learn that technology must continually be re-evaluated in light of changes in any dynamic landscape, especially something like the Internet.

Let's say I decide to build a machine, able to take potatoes up to a certain size, cutting them into French Fries. I try hard to envision a maximum size for potatoes by studying information on potato crops and determine that 99.999% of the potatoes of are less than 10 inches long, 5 inches wide, and 5 inches deep. Thus, I develop the machine to handle potatoes up to this size knowing that once in a while, a potato will come along that will not work in the machine. Six months later, a new potato hybrid is introduced by an enterprising Agronomist. This new potato hybrid is larger than other previous potatoes. Now only 85% of the new potatoes are less than 10 inches long, 5 inches wide, and 5 inches deep. The new machine I built works some of the time, but it can not dynamically adapt to the change in the potatoes. Thus, I must modify the machine to accept the larger potatoes. If I could have envisioned that potatoes might end up being larger during the life of the design, I probably would have built the machine to accept larger potatoes from the beginning.

The capabilities of technology will continue to evolve including some dynamic attributes, but so will the capabilities and ingenuity of the attacking forces. We often hear of new security technologies touted as foolproof or totally secure from an overzealous salesman who may not have done enough research to understand what is and isn't secure about the product he or she is selling. The dynamic nature of security, technology, and the ways in which technologies are assembled should prevent anyone from claiming that anything is foolproof. In some circumstances, product vendors just don't understand to a detailed enough degree, how their products will be used. Customers of products can be very creative in using products, sometimes in ways they were never meant to be used.

This goes back to our earlier discussion where we talked about the ability to envision every threat. The same is true for technology. It is not possible for a human to create a piece of technology that is able to respond to all potential changes in the landscape. Thus, in a constantly changing environment like the Internet, technology must constantly be re-evaluated in light of the changing environment. This requires considerable resources, but is necessary if we are to get the optimal value out of our technology and not rely upon it to do things that it wasn't designed to do.

In this day and age, technology has the ability, when properly applied, to delay or stifle, and in some cases prevent, people trying to obtain access to information illicitly. There is little chance that technology alone can provide the answers to protecting our information assets. One of the greatest threats to protecting sensitive information is ourselves. In security parlance, this is often referred to as Social Engineering. Social Engineering involves the education of individuals and employees to inhibit the inadvertent releasing of information that can be used to compromise systems or release sensitive data. A number of well known security breaches have been exposed through methods that have nothing to do with technology. Social Engineering attacks can cause problems for technology. How? If I can get access to legitimate information through a social engineering attack, I can often use this information to defeat the protection schemes afforded by a specific technology.

Technology is not necessarily good at protecting us against ourselves. If I get access to a corporate network and have a valid login and password to a resource on the network, it is likely that no hue and cry will be raised by the technology protecting the resource. If I use the same login and password for multiple systems, I extend the exposure that compromising this information causes. Many of us are creatures of habit and this is exactly what we do. Creatures of habit tend to perform the same activities, in the same manner, over and over. How many of us stop by the same coffee shop every day and get the same drink on the way to work? Single sign-on (SSO) technology also facilitates this in that a single login and password enables access to many systems, even though what can be done in each system may be controlled by the underlying application.

If we are not careful, the same is true for our use of Information Technology. If each user had a different user ID and password in each application, it would be much harder to penetrate multiple systems and inherent safeguards would exist to limit the damage that could be done by a single user ID and password that might get compromised. This brings us to two more of our security axioms: **No matter how much technology is leveraged to secure a resource, without the associated social engineering, it is worthless** and **unilateral reliance on technology to protect a resource is a recipe for disaster**.

One of the current places that technology fails is in Intrusion Detection Systems. Some of the current Intrusion Detection Systems of today can be fooled by what is called a Fragroute attack. Fragroute attacks break up the attack into multiple packets, minimizing the chance that the attack will be caught by the pattern matcher. Unfortunately, the host being attacked will assemble the packets in the correct order and so the essence of the attack is preserved even though it was broken up through transmission. Most IDSs (Intrusion Detection Systems) today are merely advanced pattern recognizers. This means that they look for certain types of network traffic and then respond according to set of rules when the pattern is matched. While these can be somewhat effective on very large volume attacks or attacks that use an established traffic pattern, lesser, more obscure attacks can often elude an Intrusion Detection System. Many IDSs are reactive in nature using the same technology as anti-virus programs. They compare traffic patterns to an established database of illicit traffic signatures. If the traffic signature is not recognized, the attack may proceed through unabated.

Intrusion Detection Systems can also be fooled by continually altering the signature of the attacks so that the patterns will not be matched on a regular basis. The new breed of Intrusion Detection Systems will often use statistically based information, along with pattern recognition, to determine when an attack is taking place, but these may also be fooled as well. New attacks may or may not be caught because the overall comparison is within the statistical "norm" for that server. This is not foolproof protection though since statistical anomalies can be eliminated over time by repeated transmissions of the same types of information over time, gradually eliminating the sensitivity to the errant statistical patterns. It often takes human intervention, combined with technology, to be able to catch, detect, and deter newer and more austere attacks.

Basic Training for War

In order to prepare for war, the United States spends large amounts of dollars to train it's soldiers in basic discipline, knowledge of the enemy, combat, tactics, use of weapons, etc. Given that we are also in a war, we must also prepare our security staffs for the ability to defend against the hostile elements resident on the Internet. Although costly, investing in training can at least give us a chance to protect our resources. Unfortunately, security is so dynamic that continuous training is not just a luxury, but a requirement. We can not afford to become too stagnant in our security processes, nor can we afford to be lax in keeping up to date with what is happening around us.

Training our security staffs in basic security practices is important, but we must go much further than this. Our security staffs must have broad and deep experience from which they can draw from. They need to have a working knowledge of a wide array of technology constructs including applications, infrastructure components, operating systems, networks, network services, communication protocols, routing protocols, web servers, browsers, authentication, social engineering, databases, etc. It would not be unreasonable for the most technology adept individuals in an organization to function in the role of a Security Engineer. Good security individuals are detail oriented, not taking things for granted, but investigating even the slightest anomalies in whatever they are investigating. An eagerness to learn is one of the key attributes of a good Security Engineer. This eagerness to learn enables them to operate on a level that is close to where their adversaries operate since they also are constantly learning and plying their trade. These types of Security Engineers effectively become hackers themselves, if only to understand the nature of their enemy. They are vigilant, constantly on the lookout for breakdowns in process which might lead to a newly exposed vulnerability and to constantly monitor the myriad of information sources for new vulnerabilities, closing the window on those vulnerabilities as quickly as possible. They must be solution oriented thinkers, able to devise countermeasures to new threats in a timely manner and able to effectively know where the security holes in their own environment are. Individuals that make excellent Security Engineers are a rare breed indeed.

No longer can we treat spending on training individuals in security as a necessary evil. We must train our staffs in the best security practices, testing of technologies prior to their introduction, testing of application software, configuration changes, monitoring of network traffic and system logs, educating others on social engineering attacks, developing countermeasures to attacks, monitoring vulnerabilities, analyzing risk, reviewing system and network logs, analyzing traffic patterns, etc.

We may wince at having to invest in these types of resources, but in the information age, information is our most valuable asset. It is the one thing that our enemies seek. To have a chance to protect it, we must invest heavily. Our enemies invest continually in their abilities and in technology in an organized manner to optimize their chances of success and we must be just as diligent. By maintaining a highly trained and skilled security staff, we begin to make the transition to proactive security management, moving away from many of our existing reactive security management practices.

Training does not automatically ensure that our information resources are protected, it only assists in the process. Unfortunately, there are a number of things that prevent us from totally protecting our resources.

As humans, we do make errors and with the ability of our enemies to attack us programmatically and at high volumes (literally hundreds of thousands of attacks per hour), it only takes a single error to be able to compromise a system. We have also discussed elsewhere in this book that there is a time between when vulnerabilities are exposed by either a hacker or another organization, when a patch is available, and when we actually apply this patch. The time we have direct control over is the time between when a patch is available and the time that we apply it to our systems and we need to minimize this time as best as we are able.

As we as Information Technologists mature in our handling of newly identified vulnerabilities, we will eventually get to a point where the time between when we identify a new vulnerability and the time a patch is available will go to zero so long as we do not prematurely publish the vulnerability without a patch being available. However, we must still deal with the ever increasing ability of our enemies. They can still scout and identify new vulnerabilities which they can exploit before we find out about them. This is just one more reason that security must move from a reactive posture to a proactive posture – to identify vulnerabilities before our adversaries and to close them before they can find them and exploit them. We must train our security experts to think and act upon security in a proactive manner. We can not simply wait around for vulnerabilities to be found by someone else and then wait for a patch. Our staffs must be trained to proactively discover flaws within new technologies and upgrades to products, eliminating as many of the windows of vulnerability as possible.

The Handicaps

When we wage this one dimensional war, we often wage it with significant handicaps. These handicaps significantly hamper our ability to be successful at waging the war. We wage the war with so many handicaps stacked against us, we don't even realize that our chance of successfully winning the war is greatly diminished. Some of the handicaps we play with are obvious and companies are generally aggressive about addressing those handicaps, but handicaps also exist that we don't even consider. While it is probably possible to overcome most of the handicaps through the application of a large cadre of dedicated and highly trained security personnel, the question will always remain: is it wise to try to wager the war on a battlefield where we are at a constant and distinct disadvantage from both success and economic perspectives?

Companies that have not yet had their security breached, often assume that this means that they have an adequate security strategy in place. This of course is a false sense of security and may lead to not implementing a continuously improving security strategy. Security is somewhat like investing. New information requires an adjustment to an existing strategy. The same is true with security. New information about different technologies, attacks, vulnerabilities, applications, etc. can all require a re-factoring of the existing security strategy. As new applications are deployed and new technologies are introduced, new opportunities for compromising existing security strategies are always emerging.

There can be a large number of reasons why the particular security strategy may not have been exploited to date. The obvious one is that an aggressive security strategy is in place, protecting key assets and critical infrastructure of an organization. The second is that criminal elements seeking financial gain have not yet identified a company's assets or information as a prime target. There are literally 10's of 1000's of companies now doing business on the Internet and while high value targets attract the majority of attention, it is just a matter of time before criminal elements begin extending their spheres of attention.

Often companies connecting to a public network will attempt to safeguard their systems with technology, assuming they are secure. Protecting the front door access to the Internet is often insufficient. Why? Because there can be large numbers of additional vulnerabilities that exist besides the front door access to the Internet. There are many other ways to get into an organization's internal networks including dial up modems, a business partners network, separate Internet connections, or access through a third party provider.

Each of these indirect connections to a public network represents another potential vulnerability. The problem with these potential vulnerabilities is that they are assumed vulnerabilities. That is if one of your trusted partners fails to maintain at least the same level of security that you do, you could be compromised without even knowing it at least for a subset of business functions. Similarly, if you employ a lower security posture than your partners, you could be compromising their security posture since they may treat you as a trusted partner. In addition, you have no knowledge of changes within their environment so that even if at some given point in time, you believe you are assured of a secure environment, the next change that your partner makes without informing you could compromise your security. This is the dynamic nature of trying to secure environments.

Lack of Qualified Personnel

One of the greatest handicaps that we have in trying to protect our sensitive resources is the lack of qualified security personnel to protect those resources. John Schwarz, president of Symantec, recently estimated the number of unfilled security jobs in the United States to be at 75,000. Even if these could be filled, most could not be filled with individuals competent to handle security in a distributed environment. In times past, security experts could be fairly one dimensional in nature, protecting a subset of the resources. Now that we rely on distributed systems and networks, there are so many different avenues to protect that our security people need to be multi-dimensional. They need to understand networks, traffic patterns, network protocols, physical security, social engineering, operating systems, network services, routing protocols, operating system services, directory services, databases, web servers, to name a few.

In addition, we can no longer let subject matter experts autonomously maintain security for a facet of an organization's resources. Thus, allowing a DBA (Database Administrator) to assume full responsibility for security of a database can be problematic. If the DBA assumes that all information inside a firewall is protected, the DBA may not secure the database adequately, although many DBAs may not be aware of perimeter level security constructs. In addition, the DBA may allow a vendor who provides a third party database package to have access to the DBMS (Database Management System). Without the proper considerations, this can provide an opening to internal data, especially if the database login is passed to the Database Management System through clear text. This brings us to another of our security axioms: **Effective security models can not be implemented as a series of uncoordinated activities, regardless of the experience of the constituent members.**

An effective security model requires the coordination of a variety of personnel who have expertise in a wide variety of areas headed up by individuals who have a wide breadth of experience in security matters and the vision to foresee potential threats. The development of an Enterprise Security Model is not a simplistic activity. Many organizations have multiple ingresses and egresses than can be exploited. Effective security models are implemented in multiple layers. Not only does this type of model provide some protection if a single layer fails, but it also discourages those that would compromise your resources since multiple layers of security often require a variety of different skills to penetrate multiple layers. The chance of multiple layers failing at the same time should be remote if the layered security model is designed well. In order to effectively develop a robust Enterprise Security Model, a variety of different skills are required, but the overall model requires coordination of the layers so they operate in concert, frustrating simplistic attacks with regularity. In many cases, security models are implemented in organizations by different groups of individuals without any real coordination. Thus, a DBA may assume that the Internet connection is secured, so lower levels of database security are not effectively implemented. Consider for a moment the number of different places that some form of security is implemented within an organization. You will likely come up with quite a few. All is takes is a few of those groups to assume that others are securing their resources correctly and they may not exercise the due diligence to effectively secure the resources they control.

One of the simplest examples that development of an organizational security model can not be done in a distributed manner is the case of dial up modems. If we look at the business drivers for different business units within an organization, an autonomous business unit may determine that convenience will take precedence over security.

Users in a department may determine that they would like to dial in to their own computer to complete some segment of work when they are offsite so they connect their existing computers to a phone line. Their plan is to dial into their desktop computers once they reach home. Many users are sophisticated enough to be able to setup this type of dial up capability and too often there is insufficient security protocols to prevent this from happening. If there is no modem in the computer, the business unit often has the resources to purchase and install a modem. This is a simple example where lack of control over a phone line in the voice system infrastructure can override even a vast array of security devices designed to protect the "front door".

These types of things happen in many organizations, especially if the Information Technology management group is perceived to be slow or unresponsive. When something like this happens, no matter how much security is layered over the organization's Internet connections, a simple dial up modem on a desktop computer can effectively bypass all that layered security. Thus, security models must be developed in an enterprise wide manner, communicated enterprise wide, and enforced enterprise wide. No single business unit can be allowed to "opt out" of the security model, unless they are treated as a completely separate company for purposes of security.

While most security personnel are diligent, it only takes a couple of mistakes on the part of security personnel to expose weaknesses in our defenses. We have already discussed the dynamic nature of the war. There are so many changes that continue to occur in the deployment of technology and its management, that we must continually re-educate our security personnel. New attacks, new viruses, and new vulnerabilities are found on a daily basis. While not all of these will affect all organizations, a methodical process for investigating, reviewing, and responding to these changes should be developed.

Also key to having qualified personnel securing corporate resources is the ability of the senior management team to understand the threats and to effectively analyze the risk in combating those threats. Many executives within an organization feel that level of security understanding is too complex and this is probably a reasonable assertion given their other responsibilities. In addition, few humans evaluate risk well. They assess and respond to risk subjectively as opposed to objectively. Few take the time to develop or use an established risk management methodology that identifies the chance of the risk being realized, understanding the potential liabilities incurred with the risk being realized, and then objectively evaluating alternative countermeasures. As we discussed above, it becomes effectively impossible to consider all possible attacks, but even for those attacks that we can envision, people tend to ineffectively assess and deal with those risks.

The Internet, by its very nature complicates the assessment even further. If you go to a Sportsbook and want to wager on a football game where you have some information about the teams playing, you can make a subjective decision about the risk versus reward. Even though the risk verses reward analysis you performed is a subjective one, (remember you didn't go through the process of listing all of the information factors that led to your decision and their probable impacts on the outcome) you did use some information to make the decision. This, under normal circumstances, may give you a slight edge in realizing the reward of a winning sports wager. The problem with the Internet is that no one really understands the true and complete nature of the threat on the Internet. We don't know how many people are planning to attack us, how they are planning to attack us, what resources they have, where they are, when they will attack, or what they will use to attack. It's all just one big fog that we attribute the word "threats" to.

Suppose that I informed you that China was investing more than $1,000,000,000 per year in training a staff of hackers and equipment that were designed to break into Linux systems and their applications. If you were under the impression that China was investing only $100,000 per year, you would probably assess the risk to be greater based upon the information I provided you, especially if you were banking heavily on Linux as a strategic technology within your organization. You might react by changing technology bases, ramping up your security budget, or making some other changes in response to the information. Unfortunately, since we don't have this type of information, it is hard to assess the risk and if we can't assess the risk, we don't know how to effectively prepare to address the risk.

No matter how much we would like to be able to assess the threat from the hostile entities that roam the Internet, we are unable. We simply don't know enough about how many entities there are, what entities they are, what their agenda's are, how much they are investing, where they are located, or what technologies that have access to. Thus, even if we used a risk methodology to evaluate and respond to risk, we will always find it difficult to identify the nature of the risk and the chance the risk will be realized in the Internet environment. All we can do is attempt to continue to evolve all of our security practices, hoping that they are sufficient to the task. Unfortunately, even the best organizations will be at risk. What will happen to those who treat security as a necessary evil?

One thing can be understood relatively easily by virtually all individuals and that is access. It is clearly easier to compromise a resource if there is the potential for access. Thus, one of the ways to minimize risk is to not operate in an uncontrolled environment where access to the basic network infrastructure is available by all types of entities.

Even if we pursue reasonably staffed, highly knowledgeable security teams in defense of our resources, this can be costly. There is a critical mass of experienced security experts. Very few have the depth of knowledge to effectively secure a resource on all fronts. Consideration of turning over these duties to a more experienced staff, one which can invest in continual training, one whose sole purpose is to protect informational resources, networks, systems, etc is reasonable. We must determine whether trying to secure our resources is really a core competency for us. This is one of the places that we often make mistakes in that we invest in meager security staffs because that really isn't a competency within our companies and yet we need to protect our resources.

Consider for a moment a manufacturing company. Although the manufacturing company builds certain items, the items still have to be delivered to customers. In many cases, a manufacturing company will not undertake delivery of the items that it manufacturers itself, preferring to rely on a company that has more experience in delivery logistics. The argument could of course be made that delivery of the manufactured items is key to the profit of the manufacturing company. While this is true, delivery is not normally a core competency of a manufacturing organization. Thus, it makes sense, from both a customer service and a financial standpoint to rely on a company whose core competency is delivery.

The same is true with security. Most companies, even though they have security staffs, do not invest enough to maintain a core competency in security. There is such a high cost to keep a sufficient security staff up to date and to proactively address security issues through change control, testing, technology evaluation, etc. that it can be more cost effective to let those with a true security competency handle the responsibilities, especially if they elect to provide a secured landscape on which commerce can be conducted. Later on, we shall contrast the costs, demonstrating the economies of scale, in using a dedicated security resource as opposed to trying to implement the same security functions across a variety of different companies.

Cost of the War

One of the least addressed aspects of this war is the ever escalating cost to wage it. It is almost unbelievable what we put up with as technology leaders in terms of the costs of trying to secure our information and resources. Some of the recent worms and viruses have taken huge amounts of resources to combat and have caused considerable damage to machines and to business. Many of us know that sinking feeling when a new virus is announced, not knowing if our security measures are sufficient to the task of protecting our information and resources. Is the protection we have in place sufficient, or will this be the one that gets through? Remember, that it takes only one person in an organization to activate a virus internally if it makes it past any perimeter security.

Consider for a moment the costs of some of the recent viruses and worms. The costs to combat the Love Bug virus alone are expected to reach 10 billion dollars worldwide. Computer virus experts estimated the total cost from virus losses to be about 60 billion dollars in 2003. This is $10 for each man, woman, and child on the face of the earth. This is made up of actual damages, potential loss of business, personnel required to inhibit spreading of the virus, and repairs of infected machines. These costs do not cover any direct hacking costs, but only costs directly associated with worms and viruses. The estimate is expected to increase again in 2004 with no reason to believe that it will subside anytime soon. Unfortunately, there is no way that reactive software (e.g. anti-virus software) can stop any of these types of software from doing the initial damage; they can only stop secondary infections.

When you add in all of the additional resources (people, technology, software) that has already been invested in before a company is affected by a virus, the cost of trying to secure these resources on an open network are unimaginably high. To some degree, we are in a no win situation. If we try to combat the viruses on a case by case basis, we risk being profoundly affected by a very powerful virus which can completely disrupt business operations for a very long time, perhaps even requiring activation of business continuance plans. If we are more proactive about trying to secure our resources, we still have considerable costs as we expand the size and capabilities of our internal security staffs and purchase new technologies to protect our resources and information, yet even undertaking these proactive activities come with no guarantees.

Security by Committee

Trying to secure any type of resource, when done by a group of diverse individuals can be problematic in nature. In general, the process is simply too slow and too bureaucratic to be truly effective. When a standard finally does get approved, it can already be out of date. We should reconsider our current methods of working through the development of standards and convene small groups of the best and brightest minds with these types of activities as their highest priority, in effect sequestering them. Often, because they operate behind the time curve, they end up accepting technologies and standards that have either general interest or perhaps a large market share. The government itself may not be the right entity to do this given their poor record of securing their own systems, but perhaps a small group, with government clout behind them would serve as well.

A Site Security Handbook has been published by the IETF (Internet Engineering Task Force) documenting many best practices, but many of which are general in nature. For example, it discusses distributing the minimum amount of security policy information necessary to improve security, but to balance it with the needs of the job. Thus, it is left to site specific decisions. It also addresses the continuous improvement of security policies after weaknesses are found although it offers little in terms of determining how to find the weaknesses. There are other committees that have produced similar documents including NIST (National Institute of Standards and Technology), and GASSP (Generally Accepted System Security Principles).

Committees are fine when time is not of the essence. A committee decision making structure, especially if it is a large committee, can result in a very long and drawn out process. Schedules have to be juggled, individual agendas have to be reconciled, information has to be distributed and reviewed, etc. This can be even more problematic if the committee relies on members whose sole responsibility is not the committee activities. Often, we select members because of their experience, their knowledge, or their wisdom to serve on committees. The problem is that is most cases, these individuals also have other priorities in their life whether they be jobs, responsibilities, families, etc. Each of these things detracts from the time spent working on the critical issues and there are many critical issues to address.

When you are in a war, especially a cyber war, time is always of the essence. How many wars have been won or lost on account of timing? Timing is everything in war. One of the single greatest problems with our current vulnerability warning system is that it sets in motion, a time oriented race. Once vulnerabilities have been uncovered, they are often published on a web-site accessible to both friendly and hostile elements alike. Then it becomes incumbent on the users of the technology specified in the CERT to see how quickly they can fix their systems closing the vulnerability. At the same time that the friendly elements are notified, so are the hostile elements. If they have a profile of the technologies that a company uses, (and many dedicated and organized hacker or criminal elements would have access to this) they may begin to be able to launch attacks immediately. If code development is required, they can begin to work on this immediately, establishing a race condition between the hacker and the overloaded security person trying to get a patch to cover the vulnerability. Who wins the race is dependent on the vigilance, the determination, and the resources that each of the entities applies to the challenge at hand.

Often, standardization committee's suffer from the same thing that others do – namely that security is not a high priority within the things they are trying to standardize. A considerable amount of work as been poured into the standardization of web services, but the biggest obstacle to more widespread acceptance remains security related issues. Peter Judge in a recent article on ZDNet, observed that "Encrypted web services can tunnel in and out of the corporation – exposing them to hackers. So it is not surprising to see the web services world working on security." This demonstrable example depicts how even in a committee setting, security tends to be a second thought. Security must be the priority in our standardization efforts, not some afterthought that may be clumsily bolted on at a later date. Since committees have often shown the inability to consider security first, it may be required that the Government mandates such consideration. Web services could be particularly vulnerable to hacking if not properly secured. Entities could deplete company inventories, sending them to false locations using false payment mechanisms.

Security by committee within an organization can also be risky, especially if the committee members are not "experts" in their own right or if the members have a specific agenda. Effective security models must be communicated and implemented enterprise wide, but methodical development using a small group initially and then expanding the group over time might be an effective method. If members that participate in the process of trying to develop a security model do not understand or believe in the nature of the threats, this can be a significant liability and will often inhibit real progress. In addition, care must be taken to eliminate all potential political agenda's that may influence the final procedures.

In addition, many entities not have been willing to defer to security experts insisting that their knowledge of security is sufficient. It is literally unbelievable how complex it is to secure some types of resources. Depending on how the management dynamics in the organization function, they may rally others to their side effectively blocking the progress in developing the security model or attempting to assert values that are not security oriented. When values, which do not have a security first mindset, are inserted into the process, all sorts of compromises begin to occur and we see the evolution of security models that emphasize convenience over security. Examples of these types of compromises include simple passwords, dial in access to local systems, reductions in processes designed to identify people, etc. We must be on guard to retain the pure nature of the security process and not let it be hijacked through committee structures.

The Price of Openness

There are several attempts underway, even as this book is being written, to develop standardized security models. This violates one of the fundamental premises of security - security models and protocols should not be discussed openly or completely standardized. Best security practices should be developed behind closed doors and then carefully distributed to a limited number of individuals, on a need to know basis, whose identities are a known commodity. We hold security conferences where we openly discuss vulnerabilities, new security technologies, standardized ways of securing resources, enhancing social engineering skills, etc. Doubtless, at least some of the members of the audience, especially at the larger conferences, are collecting information intent on using the information in the pursuit of hostile activities.

When we open up information in this manner we give our enemies an edge. Remember, the first step in defeating a security strategy is to know of its existence and its nature. By the way, this extends to things other than information security as well. For example, sometimes precious artifacts are protected by light beams which activate alarms when they are crossed. The light beams are normally invisible so that people who would like to get access to the artifacts illegally can be caught. Once a criminal can tell the location of the beams however, they have a basis of beginning to find their way though them without tripping the alarms. Security professionals often advocate the use of authentication (validation of who we are dealing with) in securing resources, yet when we discuss security issues and divulge information on our security measures, we often don't even know whom we are divulging details to. Rethinking our whole approach about being open about security is necessary if we are going to have a chance to protect our resources. Security is not something that one boasts about. Those that are most successful at security are those who are more interested in keeping the nature of their security secure rather than trumpeting details for all to see.

Security models on an open network should not be the same for all entities using the Internet, because once a method of circumventing one is found, similar models can be at risk. We have learned this through other, non-computer related activities. One of the greatest failures of the nuclear era is how easy it is to get access to information on how to build a nuclear device off the Internet. Even as we speak, Iran is being investigated for the development of nuclear weapons and North Korea is manufacturing weapons grade plutonium and building nuclear weapons. Did they get the information off the Internet, or possibly through our educational system? No one will probably ever know, but the Internet is such a vast repository of information, that this is a possible scenario.

While this author understands that it is not realistic to keep all information private, there is a vast difference between actively publishing information and keeping the same information secure with the potential that it will eventually come out over time. When information is assimilated over time, it may or may not be accurate by the time all of the information is generally accessible going back to our concept of Information Half Life. For the same reason that our military does not publish their war plans on the Internet prior to a planned engagement, security needs to be addressed in a much more closed environment.

Have we ever considered the cost of openness? I know that when I speak at an engagement, I share very few specific details as to the security protocols of our organization, our technologies, or our usage of information. I don't want to give individuals with criminal intent any more information than they already have about our environment and I don't always know the exact makeup of the audience. This author would contend that our "openness" makes us vulnerable, but more than that, it dramatically increases the cost to protect ourselves. It costs us time and money to combat criminal elements breaking into our systems, yet many of these same entities got the information to attack us off the Internet.

Of the seven new vulnerabilities per day that were exposed in 2002, approximately 60% of the vulnerabilities could be easily exploited because they either required no code to exploit them or the code to exploit the vulnerabilities was easily available. Often, some of the institutions that are designed to protect us expose us to danger. It is easy to look up for example, specific vulnerabilities posted at the CERT site (www.cert.org). While it would be difficult to estimate how many of the United States Government entities and companies had not addressed these vulnerabilities yet, we can be sure the number is probably rather striking. The Slammer virus which attacked in January 2003 could have been addressed more than six months earlier since that was the time the patch for that vulnerability was released. This time lag makes us vulnerable.

Another unrelated example involved TV stations showing how weapons can be smuggled through security checkpoints at airports. While they may not recognize it, reporting on something like this, puts us all at risk. Since it takes time to close the potential holes in the airline security, there is a window of vulnerability which exists between the time the security holes are exposed and when they are closed. If the TV stations were truly interested in our security instead of sensationalism, they would have been better reporting it, in private, to the FAA and then perhaps reporting on it to the public after the holes were closed. In addition, the report on the vulnerabilities in airport security can get others thinking about how a slight mutation of the vulnerability that was exploited might not end up getting caught.

Let's now look at some of the costs of openness in society. Listed below is a table of some of the places that we are open, the potential results of this "openness" and the potential costs incurred. While most of the examples below are real world examples, these are just a few of the many that exist. We could talk about many others places that we are open where we ought to be more discreet.

Area of "Openness"	Result of "Openness"	Costs Incurred
Nuclear Technology	More countries having access to nuclear technology; inherently more unstable world; potential for reactor scrams releasing radiation for less experienced users	Requires additional military diligence and spending; foreign aid to "control" these elements often needed; potential for loss of lives and environmental costs
Sex	Younger and younger children becoming pregnant	Health system costs; some individuals not fully educated; higher ratio of welfare cases; higher incidence of child abuse requiring family services intervention; more abortions; higher costs for medical care for sexually transmitted diseases
Information System Vulnerabilities	Hackers have access to information to hurt us; reduces their time to launch effective attacks	Larger and more active security staffs; more technology and policies required to protect information resources; greater need to close windows of vulnerability quickly; cost of combating worms and viruses

Area of "Openness"	Result of "Openness"	Costs Incurred
Serial Murder Coverage	Copycat criminals appear	More lives lost; often requires more investigative resources due to multiple perpetrators
Airport Security Lapses	Exposes how terrorists can get access to critical areas	Provides new thinking and opportunities for terrorists wanting to destroy airplanes; requires additional security to counter evolving threats

This chart above is a rudimentary chart, but depicts some of the issues associated with being open. We as Americans take it as a right to have this level of "openness" in our society, but we must learn to ask if it is really proper to be open about everything. There is no amendment in the Constitution that guarantees our right to know everything. While there are many things that being open about and educating people around is great, it is the distinction between what is appropriate to be open with and what is not appropriate to be open with that we have not yet mastered.

Our enemies are active in many ways. We can not afford to be arrogant assuming that being open can not hurt us. Even if we are able to effectively thwart threats against us because we have been open, the cost is significantly higher than if we simply had used a little judgment in determining what is and what is not fit for public consumption. As a very excellent commercial from a large computer company touts "Cool costs me money". So does openness. You might consider some of the observations in the previous table prudish or may rationalize it in some other manner, but the reality is being open, costs money, costs resources, and in some cases, costs lives.

Now think about the costs associated with operating on an open network by way of specific examples. How much additional security technology is required? How many additional staff members have to vigilantly maintain watch for signs of attacks? How much needs to be invested in ongoing training? What is the cost when your systems are breached? It will quickly become clear, that there is a very high price to be paid for openness. Although we consider ourselves to be living in a very open society, there are a number of instances where sensitive information is protected. Why are they protected? They are protected so that damage will not be done to individuals or entities. We protect sensitive information about individuals such as their medical histories or criminal histories so that they will not be discriminated against.

We as a nation have not learned to discriminate between when openness is appropriate and when it is not appropriate. I believe that even today, there are some media outlets that would release the war plans for the United States to boost ratings, regardless of the potential cost of United States lives. Recently, a recent news correspondent was requested to leave the Iraqi theatre of war for revealing sensitive information about military plans. While this was an error in judgment by the media correspondent, it was also a lapse in judgment by the military official who failed to protect it. While striving to be an open society is not all bad, we are left with the question: What is the cost for being open? If we are unwilling to pay the cost in time, in dollars, and potentially even lives, we need to start distinguishing what information we can be open with and what we can't. Consider even the mundane Social Security Number. At one time, the Social Security Number was not a very well protected piece of information. Even today it appears on many public documents. The result? Identity theft is one of the fastest growing crimes and the Social Security Number is a facilitator of that.

With respect to securing information and critical resources, the cost is not generally in human lives, but it can be close to this. By trying to protect information traversing an open network, we run up considerable bills in training of social engineering skills, network security hardware, encryption technologies, monitoring devices, training on security technologies, enlisting a dedicated security staff, security software, authentication software, security assessments, unannounced penetration testing, etc. The list goes on and on. Even after all this is done, we still can not guarantee security. Thus, the cost of openness is very high and the price can climb even higher. We have only spoken about the costs to try to secure a resource. If the information is compromised, even after all these security measures have been implemented, now our employees and our customers are at risk. They may lose money from their bank accounts, have their identity stolen, have their credit rating ruined, or perhaps even worse.

Fairly recently, the New York Electronic Times Taskforce reported on the arrest of an individual who has been accused as an identity thief. He had amassed information on the 400 wealthiest people in America having collected such information as social security numbers, mothers' maiden names, credit card numbers, etc. He stands accused of trying to use a variety of different methods (Internet based and non-Internet based) to steal their assets. In mid 2001, the Justice Department identified identity theft as the "nation's fastest growing white collar crime" while some have called the misuse of social security numbers used for fraud as a "national crisis". Medical and criminal records are also at risk which could affect an individual profoundly.

The Endless Stream of Changes

Even if it were feasible to consider all possible threats to a piece of critical infrastructure, it still may not be possible to actually secure the critical infrastructure. Software is the pre-eminent example. Most software is not static, but dynamic in that it constantly changes, undergoing maintenance, enhancements, and upgrades. Even if we assume that at some point, an application, its infrastructure, its data, and its access points are fully secured, the application will change and so will the systems and infrastructure software (operating systems, web servers, application servers, databases, etc.) that run the application. When the infrastructure software components begin to change, new holes can appear in a previously secure infrastructure.

In order to effectively try to keep up with these changes, a number of ongoing activities are required of an Information Technology staff. The first is security monitoring. This activity includes the continuous monitoring of entities that collect and process security related events, incidents, and vulnerabilities. New incidents are reported every day and it is incumbent on an organization that is serious about security, to review these incidents and vulnerabilities and determine their applicability to their environment. Once applicability has been ascertained, it is necessary to develop timely plans to close the vulnerabilities hoping to close the vulnerability before an attack is launched against the vulnerability. In a well developed and layered security model, it is often possible to utilize another layer of technology or process to compensate for an identified vulnerability until that vulnerability can be directly addressed.

Although in today's environment, this is a reactive aspect of security and not a proactive posture, the goal is to get to the point where the companies are notified about vulnerabilities before they can be exploited by someone else. When the industry gets to a point that it notifies only authenticated individuals of vulnerabilities, the clock on exploiting those vulnerabilities will not start for the hacker at the same time. That doesn't mean that a hacker can't discover the same vulnerability, but at least we won't put additional information into their hands that they can use. What we want to do is understand the planned changes to our environment and then deal with security issues prior to their introduction into our environment. In some cases, we can delay introduction of new technologies to an environment until we research existing vulnerabilities and probe for new vulnerabilities.

Another activity required to address dynamic changes is proactive testing. Security testing of applications and their infrastructure is also a continuous activity that must be re-examined whenever a change is introduced into the environment, whether it is the installation of new software or even something as minor as a configuration change. This means that prior to the deployment of an application (or a new version of the application), the application should be fully tested along with the infrastructure that the application will run on. When a change occurs (production fixes, maintenance patches, change of web server releases, change of application server products, upgrades of operating systems, configuration changes to a router, etc.), the same set of regression tests should be run again to validate that no new security vulnerabilities have been exposed. Change control for all changes must be examined to determine whether or not there are security implications. We tend to react quickly to needed changes in our environment, but this is one place that a hasty approach can get us into trouble. Planning out and testing the impact of changes is important but takes time thereby increasing the overall cost of making changes.

A continual process of research and testing must also go on throughout the software development life cycle. This brings us to another of our key security axioms: **It is cost prohibitive to try to keep up with all of the changes in security on a daily basis that an open and dynamic network requires.** Why? Because there are so many things going on at one time that it takes a large, dedicated staff to be able to keep up with what is going on. This is the treadmill effect that we discussed earlier. We continue to reach for a fully secure network, but we will never reach it on an open network; it always remains just out of reach. While we strive for it, we spend large amount of capital to try to improve it. The whole open environment costs us money.

Consider a partial list of tasks that a potential security staff must undertake on a recurring basis. The actual number of personal required is contingent on the number of servers, number of databases, network devices, applications, employees, etc. that an organization has. Listed below are many of the tasks that need addressing on a regular basis in order to deal with the endless stream of changes.

Partial List of Tasks Needed to Attempt to Secure Corporate Resources
Monitor New Vulnerabilities
Determine Impact of New Vulnerabilities
Develop Plans Against New Vulnerabilities
Develop a Thorough Security Regression Test Suite and Keep Up To Date
Regression Test Applications for Security Vulnerabilities
Test Operating Systems and Infrastructure Software for Security Vulnerabilities
Introduce New Technologies Into Testing Lab
Test New Technologies in Testing Lab
Provide Vendor Feedback on Vulnerabilities
Setup and Configure Network Security and Monitoring Devices
Setup and Configure Security on Data Sources
Setup and Configure System Level Security
Review System Logs
Review Application Logs
Review Physical Access Patterns
Perform Code Reviews
Review Traffic Patterns, Looking for Attacks
Develop Security Policies
Provide Education on Policies and Social Engineering
Develop Social Engineering Strategies and Implement

Within your organization, how many things above are done on a daily basis? Is there anyone reviewing new security vulnerabilities discovered during holidays, weekends, or time off for the security staff? Given that there were approximately 2,500 new vulnerabilities that were exposed last year, this equates to about seven (7) per day every day of the year. Consider the size of staff it would require to constantly keep abreast of those vulnerabilities, analyzing them to determine the impact to your existing environment and then developing plans to deal with the vulnerabilities. Worse yet, there is no end in sight.

A Key Conflict of Interest

Product companies tend to be focused on the bottom line. Many vendors implement security as an after thought in their products. Given the number of security CERT's issued against products, one of two conclusions is inevitable. The first possible conclusion is that product companies do not spend enough time designing in security as the primary requirement for their products, preferring to ship them before they are really ready and recognizing the revenue at an earlier point in time rather than spending the additional time in the SDLC (Software Development Life Cycle) to build secure applications.

The second possible conclusion is that the organizations producing the products simply can not consider the wide variety of threats that can be levied against their products. Keep in mind, the product companies are generally the entities that know their products best. If they are unable to consider all the threats against their products and code to prevent them, any other organization operating the technology will be at a disadvantage in securing the same technologies since they do not have the same level of knowledge that a company developing them has. In addition, companies which use the products of vendors rarely have access to the source code so they are unable to review and compensate for security breaches, having to wait for the vulnerabilities to be exposed through an attack. Consider the rate at which new incidents are reported to CERT/CC.

Incidents Reported

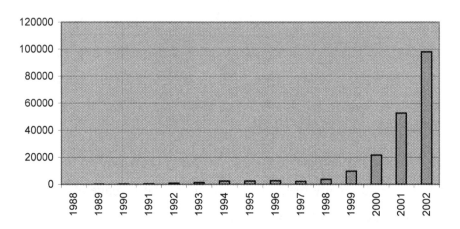

Are you aware of anything else that has grown in a similar manner to the graph above? The answer of course is the Internet. The Internet growth statistics, both from a host and individual user count have a very similar growth curve. Thus, the wider the usage of the Internet, the larger the number of incidents reported. In reality, if we were making progress in the war, the curve should be flattening out or actually declining. Also, the incidents above are exactly that – incidents, which may affect one or more computers so the actual number of computers affected by these incidents may be much higher. We would have hoped that new users would learn from earlier mistakes by others, but to at least this point in time, this does not appear to be the case.

Product companies simply do not have the ability to envision all possible attacks or the time to ensure that software that is secure. One other thing that many companies lack in order to develop secure software: knowledge. While some programmers have the experience to develop secure software, most do not. And even if they do have some knowledge, it is often dated. It would take a considerable reduction in the number of productive developer hours to keep developers current so that they can develop more secure applications and even then, there is no guarantee as to how effective that would be.

Another consideration is the way in which software is built. Software is often built out of software components. Some of these software components are developed by internal software developers and some of them are used from other sources. These sources can be other software companies or perhaps even open source repositories. Even if all of the locally developed software is developed securely, that does not guarantee a secure application or software. The other software components that are included in the overall software system may not be secure and thus vulnerabilities can be introduced into software systems. Unfortunately, there is no way to effectively ensure a secure application. Appendix A, which contains a list of vulnerabilities in software discovered over a three week period, are distributed over a wide variety of technologies.

Even though we may not be able to guarantee secure software, we can certainly improve our effort at developing it. First and foremost, security within an application must move from an afterthought to the prime consideration. Security must be designed in from the beginning. In addition, developers, for the most part, do not spend large amounts of time understanding security issues and do not go through regular training updating them on the nature of new security attacks. Many have a limited understanding of network issues that could affect their applications. Designing in security from the beginning also makes very good business sense. Several well known studies have documented the value of correcting software problems as early as possible in the SDLC (Software Development Life Cycle). It is always more economical and more effective to try to catch any developmental issues as early as possible within the Life Cycle. Trying to make major modifications later in the Life Cycle costs more and can be very hard to do in an effective manner. When we try to implement security and security fixes as an afterthought, we can end up introducing new vulnerabilities during our attempt to improve the overall security of the base software.

Security testing, an art that is not practiced consistently enough in the development of software, must evolve into a prime development consideration. Security testing often requires white box testing. White box testing involves more than simply looking at inputs and output, instead requiring the internals of the software to be carefully examined. Many of the security holes that exist in applications are side effects that appear inadvertently and were not part of the original functionality. They can arise through a wide array of application design issues including inter-component communications, integration, file usage, database storage, authentication and authorization, lack of pre-emptive validation, and insufficient exception handling routines.

Error handling code is one of the most important issues to deal with in the testing world and too many testers focus on the "happy path". The happy path is the way an application executes under ideal conditions without the need to resort to exception processing code. Hacking is all about detecting and exploiting exceptions. White box testing requires us to understand <u>how</u> something happens in an application not just <u>what</u> happens. Our testing processes need to evolve beyond just functionality testing and a full security testing process needs to be used where the main goal of the security testing is to fully exercise the exception handling code and to try to mangle the application into unexpected behavior, exposing un-handled exceptions.

An example of why this can be important is to consider an application, written in HTML (Hyper Text Markup Language – the language used in many web pages) which uses hidden fields. These hidden fields, which are not directly viewable in the browser, can be viewed by looking at the source of the HTML. Since developers at times will use these fields to keep important information that they need to send back to the web server via the HTTP Post method, finding them can be important. Once the source has been viewed, it can be saved and edited and then resubmitted to the server. If there is no validation on the information sent back to the server or the developer does not eliminate the use of "hidden" fields which the user can actually see, this opens up the possibility of performing fraudulent transactions. Similar exploitations can be executed through other methods such as the HTTP Get method when parameters are passed through the actual URL (Universal Resource Location). A very simple example of this type of URL is below:

```
http://howtochangeprice/script.cgi?xname=widget1&price=300
```

which might be edited to

```
http://howtochangeprice/script.cgi?xname=widget1&price=3
```

If there is no server side validation, we just changed the price to $1/100^{th}$ of what it was designed to be. Cookies can present the same types of problems in that they are data, accessible to an end user, which can be modified. Sufficient testing can catch these vulnerabilities, but using these types of techniques is risky. Who knows what other sensitive information is passed in this manner on some web sites?

Regression testing is a method often used in software development to ensure that bugs fixed in prior releases do not appear in later code lines. Regression testing generally utilizes a representative sample of the total test cases to predict whether or not the bugs fixed in a previously released version of the software remain fixed. Depending on the nature of the items being tested, a full regression test suite may be required. Something similar to this might be valuable in the security world. Security testing is one of those areas that would require that the regression test suite would have to be comprehensive in nature because a single exposure could result in large liabilities.

This still does not allow us to necessarily envision all potential future threats against the software, but it can make sure that we don't reopen holes in software products that had previously been closed. In addition, the approach is proactive in nature, attempting to close vulnerabilities before deployment rather than deploying and reacting to vulnerabilities after the application is in production. Regression testing can also be applied to technologies as well so that vulnerabilities can be tested for in operating systems, web servers and the like. Regression testing doesn't happen to the degree that it needs to in most organizations. A recent security patch sent out by a vendor, designed to fix a series of security holes actually ended up locking users out of web sites and preventing the use of email. Clearly this type of error was not caught because of either insufficient regression testing or no regression testing at all.

The world of software is getting more and more complex and this translates to the potential for more and more vulnerabilities. As the complexity of software continues to increases, it becomes more and more difficult to effectively test all of the software to ensure that there are no glaring security vulnerabilities within the software. While the trend towards more and more complex software is not likely to change anytime soon, we can change our posture towards security. Vendors can enforce security testing regimens to ensure that they are making every effort to secure their products, but this of course will require sacrificing (or rather delaying) some profits. While vendors need to undertake this level of security testing, each individual organization needs to do the same thing with both the software that they develop and the software that they use which may have been built internally or by others.

While product organizations may complain as to the cost of developing secure software, the fact that it has been shown to cost more to correct flaws in the later stages of product development and post-deployment should motivate them to ship more secure products initially rather than spend large amounts of resources in trying to act reactively to newly identified vulnerabilities. Add to this the potential for product liability suits to emerge and it becomes clear that we must move towards the development of more secure software from the start.

It was once said, that if builders built buildings the way that programmers develop programs, the first woodpecker on the scene would destroy civilization. To some degree this is true. Each time we patch an operating system, a web server, or a similar component, we open up the possibility that for each hole that we close, we open two more. Continually patching software is a risky proposition in itself. Once again we see the challenge in trying to secure that which can not be fully secured. Although a popular answer to this is to completely redesign our software base with security being the prime requirement, this is not realistic given the inability to develop perfectly secure software. A more realistic approach is to move the usage of these technologies to a place where there are many less hostile threats.

We have seen vendors haggle amongst each other declaring their products to be secure. They do this because security is on the minds of executives and this attracts business. We have seen the campaigns from vendors starting that they "secure the Net" or that their products are "unbreakable". Unfortunately, this is false bravado on two fronts. First and foremost, it is arrogant to assume that any one company can secure the Internet or that they have been able to envision every possible threat against their products and then develop foolproof countermeasures. By definition, a layered security model relies on multiple technologies and multiple processes so that if there is a failure in one layer of the model, there is at least a chance that successive layers will compensate. What product companies should be stating is that they have a commitment to securing their products, not that they have developed a series of totally secure products.

Second, even if these and other vendors were able to develop products which operated in a secure manner in a pristine environment, that

159

doesn't translate to real world experience at customer sites. Customers, often inadvertently, open up holes in reasonably secure products through a variety of different actions including the erroneous configuring of software.

This implication may be very serious for companies which have large install bases of software. While executives in large software development companies undoubtedly direct internal resources to pursue the development of secure applications and infrastructure software, the approach is a complicated one, especially if the strategy is to try to secure an existing software base. Trying to bolt on security to millions of lines of software as an afterthought is a very difficult thing to do. Often companies will take an alternative path and try to be reasonably reactive to security problems that are uncovered with their software. Those companies which have poor security track records and did not adequately design security into their products from the start will likely have to start over to have any significant measure of effectiveness. The cost of this approach is high, but this may be the only way to succeed in securing applications and infrastructure. Even with this approach, there is the potential for failure if the organization developing the software is not prepared to develop secure software. Software is getting more and more complicated everyday, with more lines of code and more functionality. As software gets more and more complicated, the problem of securing the code also becomes more complicated.

In the future, vendors may very well be required to demonstrate, perhaps in a court of law, their diligence in developing secure products. They may be asked to produce staff training plans, documented evidence of execution of the training plans, initial product security architectures and designs, results of code reviews throughout the development lifecycle, comprehensive security testing suites and test plans, and similar security measures. It will be necessary to demonstrate a consistent pattern of attempting to develop secure software and hardware.

In addition, vendors will need to improve their documentation stating specifically how and where a product is designed to be used to avoid the inadvertent use of the technology in the wrong places. The documentation must provide more descriptive information on configuration, establishing what each configuration parameter can do with respect to security. Thus, as with the companies using the products, it will fall to the vendor to demonstrate the routine exercising of best security practices.

If vendors are unable to show this level of diligence, they may become liable for damage done using their products. If companies that developed products were held legally liable for deficiencies in security this might affect the product delivery cycles and thus revenue streams. I believe that in time, this is exactly what will happen. No longer will vendors be able to develop insecure software with impunity. To this date, prosecution of hardware and software vendors producing insecure products, through negligence, has not been a routine activity. While no vendor can effectively guarantee secure software (and for the most part hardware), the time is coming when vendors will be required to demonstrate, empirically and academically, their due diligence in designing and implementing secure hardware and software. Yet because we choose to operate in an uncontrolled environment, the flaws in the development of products will continue to be exploited.

In addition to product companies, we now have to deal with the ever-evolving world of Open Source. Open Source software has a high rate of security advisories opened against them. Of the 29 advisories issued by CERT through October 2002, 16 of them were issued against Linux or Open Source products. Advisories are generally issued in smaller numbers than vulnerabilities because of the impact of issue or the install base.

A recent article in Computerworld addressed the discovery of a number of flaws in the Linux operating system. The article reported that many users of the operating system were unfazed by the vulnerabilities. These vulnerabilities were severe in that they could allow an attacker to gain administrative control of the system, allowing them to run a wide array of additional attacks. This demonstrates the nature of open source software in that while it may be initially more cost effective than proprietary software, it is not necessarily more secure, nor is it necessarily cheaper in the long run.

Before the acquisition of any technology, it is important to consider the post installation costs of trying to keep open source components secure.

Developers contributing to Open Source repositories are often more interested in functionality than security. Many of the developers contributing to Open Source repositories have no formal training in application level security. Since many of these components make their way into Internet applications, they can introduce with them a series of vulnerabilities that could compromise the remainder of the application. There are so many levels on which to attack an application that must be considered. Buffer overflows, un-validated parameters, invalid access control, broken session and account management, weak authentication, scripting flaws, command injection flows, invalid error and exception handling routines, lack of sufficient administration security, and application configuration miscues can all lead to insecure applications. New open source projects such as OWASP (Open Web Application Security Project) are underway and should be applauded by all. Unfortunately, penetration of this type of project among the millions of developers will occur slowly and will take many years to reach the masses. Even so, there is no guarantee that this type of project will be comprehensive.

As discussed earlier, the dynamic nature of the war does not allow a limited training effort to hold much value. Application development security training (and other training) must be continuous to deal with the constant changes in languages, products, networks, etc. This brings us to another of our security axioms: **The dynamic nature of an open network requires an ongoing commitment to training for all parties involved in designing, developing, deploying, and configuring applications.** This covers developers, architects, systems administrators, database administrators, network engineers, etc. Given the rate at which technology changes, it is not unreasonable to have to spend a minimum of 4-5 weeks per year per individual in security related training, not to mention the associated social engineering issues which also require education. Even with this level of commitment to training, there will undoubtedly still be plenty of vulnerabilities to exploit.

Liability

One of the trickiest issues to deal with is liability. Ask any of your vendors if they would be willing to cover all losses to your company and your customers if you adopted their security recommendations and you are likely to get a hearty chuckle. Software (and hardware for that matter) is not warranted against security breaches. It is not warranted to be secure for a number of reasons including the inability to foresee and counter every threat. In addition, the overall cost of trying to build a highly secure product is considerable. Many vendors do not allocate the necessary amounts of resources (both time and money) to address security issues within their products, nor do many design with security as one of the primary considerations. Even if product vendors did offer a guarantee, it would likely be limited to the price paid for their product. Very few vendors, especially if they have widely deployed products, have coffers deep enough to cover inherited liabilities such as fraud caused by the compromise of millions of credit card numbers and associated personal information.

In reality, software and hardware companies don't want to be forced to ship only secure products. Based on our earlier discussion, it is impossible to develop a totally secure product, but even if a company could do this, they would probably rather ship an insecure product quicker than a totally secure product at a much later date. Companies are in business for profit and any additional costs incurred in the development of the product, as well as delays in shipping products, translates directly to the bottom line. This does not excuse us from doing everything we can to protect information; it simply means that we can not look to companies to deliver secure solutions.

With new laws that are beginning to appear holding company officials responsible for their company's results, a large security exposure could end up being more than embarrassing. To this point, we have not seen consumers or employees hold companies responsible for not protecting their information, but this is probably likely in the future. Even if a company can prove itself diligent in protecting the information of others, the legal costs from an onslaught of lawsuits can be expensive. When large amounts of information are compromised, this opens the door for class action lawsuits with the potential for larger penalties. Once this level of negative publicity has been reached, even more negative consequences will begin to surface for the organization. Consumers may turn away from doing business with a company if it comes with significant risk of personal information being compromised. Just ask a person how much work there is in trying to repair credit records after they have been the victim of identity theft.

Companies can also incur liabilities because their servers, if improperly secured, are used as the platform from which an attack is launched against other entities. This is referred to as downstream liability. Cyber crime prosecutor Marc Zwillinger, of the United States Department of Justice, indicated that companies or other entities owning servers, from which attacks are launched, could be prosecuted. This was an important evolution since most liabilities result from a breach of contract. Although there would not necessarily be a contract between the entity and the victim, the entity could be held liable of committing a tort through their negligence. This statement implies that enterprises can be held liable if they can be shown be negligent in addressing their security and over time the definition of security negligence will likely be expanded.

Another threat to the business arises when we fail to protect the information of our customers adequately. If a company demonstrates that it is unable to protect its informational resources either in storage or through transmission, clients may begin to be wary of doing business with that company in any but traditional means, limiting the value of their e-commerce strategy and hurting their competitiveness in the electronic marketplace.

As we have seen above, a large number of activities need to be undertaken to effectively secure resources. Each day that passes sees us come closer to true organizational liability for not adequately protecting information and information resources. In addition, depending on the nature of the attack, and the damage done, there could literally be hundreds of millions of dollars in liabilities created for an entity, especially if significant loss of life occurs. Even now, there are lawsuits filed against the airlines for negligence in the September 11th attacks. In a similar opinion by Scott Zimmerman of the SEI (Software Engineering Institute) at CMU (Carnegie Mellon University), he declares that owners of Information Technology assets have a duty to keep systems secure and prevent them from being used by others to do harm. The same has existed in the past for similar examples. If a parent allows a gun to get into the hands of a child and that child does harm with it, the parents can be held liable. This definition of this type of liability will evolve over time as will the basis of diligence and negligence, but these will always be subjective areas. What I might consider negligent on your part, you may consider diligent, thus we will be left with the question: how much is enough? While this is difficult to answer, there will be many things that I believe will be used as a ruler to determine whether or not due diligence is being practiced. A summary list of these items include the development and implementation of an enterprise wide security model, keeping current with maintenance on different software, a robust training process, a well defined security testing and new technology introduction process, separation of responsibilities, vulnerability review, identification, and remediation, and so forth. This list could go on from here, but one thing is clear: it will be up to an organization to investigate what is required to be diligent and to assume responsibility for protecting the information and systems in their charge. These types of activities are going to require significant staffs – either internal security staffs or external consulting resources.

Today, a consumer may share their personal information across so many transactions that it may be impossible to prove where the information was compromised. This does not allow us to skirt the liability issue however, since a criminal that is apprehended can confess to both the stealing of the information and often, the usage of the stolen information. If they will not confess, often the information on their hacking systems can provide this information as can paper logs and other evidence.

Because I have personal concerns over this type of fraud, I tend to use several different credit cards for different sites (I do a limited amount of Internet shopping) so that if my information is compromised, I have a pretty good idea of through whom it was compromised. Also there will be both consumers and employees, who have had private information stolen, that will come forth with information that has been shared in only one place.

We already know that large groups of Americans are concerned about the privacy of their information. Some are almost at the point of being paranoid about it. Thus, if we truly want to realize the full range of benefits from electronic commerce, we are going to have to get serious about protecting people's information. Each new horror story that we hear in the media, affecting systems on the Internet, undermines our credibility in claiming that we are effectively protecting people's data. As this is written, new laws are being enacted, the first of which is in the state of California, which will force companies to reveal their security breaches to a number of different entities including consumers, the press, and others. If the company operates only as a cyber business, the clients may seek alternatives to doing business with that company by buying from other companies that provide the same or similar products, but who have a more diligent commitment in securing personal information. This introduces another of our security axioms: **Protect information resources as though they contribute directly to the bottom line, for in the future, they will.** If a company is particularly diligent, this may result in additional revenue through a consumer's awareness of your diligence in protecting their data, but even if this is not the case, protecting informational resources can prevent an entity from acquiring potentially significant liabilities.

There are cyber liability policies that can be obtained, but these require companies to have an effective security policy and can be prohibitively expensive. In addition, most of these cyber liability policies protect the company's assets, not necessarily the information of their customers or their employees. As we will see a little bit later, cyber liability insurance is by no means a silver bullet. In the future it is possible that this type of insurance will either not be available at all or will at least be prohibitively expensive. Another potential is that the policies will become so restrictive that they only cover a small number of areas rendering the policies of very limited value.

Cyber liability insurance attempts to mask or compensate for the fundamental issues surrounding securing environments for electronic commerce. In a move something similar to what happens today with car or home insurance, I believe that the cyber insurance industry in time will adopt the same type of incentives that existing car and home insurers do today. This means that if you want to operate your business over a public network and communicate sensitive information where others have the potential to access it, the premiums will be higher, eventually skyrocketing. Determining an appropriate rate for a cyber liability policy can be a very tricky thing to do. This is one of the largest areas for concern within the cyber liability insurance industry. Exactly on what, should the premiums be based? It could be based purely on technology, but this is insufficient. Due to the dynamic nature of the environment they are trying to insure, there are so many different factors that come into play.

Here's a short list of factors that I developed which might be required to insure an entity and that I would be looking for. Note that the list below is neither exhaustive nor does it guarantee that the entity being reviewed will be able to protect its data, but what it does demonstrate is the awareness of the importance of security within an organization and how well an organization's security strategy has been thought out. In general, it is probably more important to ensure that an organization has a rock solid commitment to security with some flaws as opposed to a flawless point in time security with questionable commitment. If during the initial audit you find many of these things

171

missing, I would contend the risk to insure such an entity would be high.

Factor to Be Considered	Impact on the Premium
Size, experience, and training of security staff	High
Average # of vulnerabilities per time period discovered within existing technology base	High
Vulnerability review process	High
Average time between vulnerability discovery and vulnerability closure	High
Implementation of layered security architecture	High
Monitoring and analyzing tools	High
Log review and auditing procedures	High
Installed technology base	Medium
Incident response plan	High
Proactive technology investigation process	High
Commitment to security from senior officers	Very High
New technology introduction process	High
Application security testing staff and procedures	High
Network Engineer security training/abilities	Very High
Database Administrator security training/abilities	High
Developer security training/abilities	High
Security technologies in place	High
Social engineering education and policies	High
Change control procedures	Medium
Log and traffic review procedures	High
Data protection mechanisms	High
Value as a potential target	Very High
Risk analysis methodologies	High
Potential liability from exposure of sensitive information	Very High

How would the efficaciousness of some of these things be measured? They could be measured through a series of tests, interviews, unannounced ethical hacking attacks, review of social engineering policies and procedures, and so on. This is a clearly a problem. If even one of these areas is completely bypassed, it can render many of the other security measures ineffective. It takes invasive investigation, on a company by company basis, to ascertain the effectiveness of a company's security measures. Even with this in-depth investigation, the information can be very dynamic in nature. Even if the review is well done, things change. Suppose you get a favorable cyber insurance premium which may be largely based on 1-2 security superstars and the activities they undertake within your organization. What happens is they leave the company right after the insurance company's audit? How does this affect your premiums? Do you have to notify the insurance company of changes in technology, changes in staff, etc? Can the organization sustain its security posture without the superstars? What if the superstars depart the organization? How many other factors must be continually be evaluated?

If the current rate of increase of vulnerabilities exposed per time period continues, there is the potential that eventually cyber-liability insurance will cease to exist or the sphere of things that it covers will continue to be reduced. The profiling of cyber liability insurance policy customers will likely get more and more invasive over time. If a company begins to experience failures, the insurance companies will likely raise their premiums to accommodate high risk customers. I would expect that insurance companies providing cyber liability insurance will keep their customers on a short leash since a single vulnerability can result in huge losses. Eventually, high risk customers, the customers that fail to take adequate steps to protect their data, will be refused coverage. In order to keep premiums low, it will be incumbent on companies that they take every precaution to protect systems and sensitive information.

Even though there are companies issuing cyber liability policies, the art of effective risk assessment is a black art with few well defined measures of risk. Thus, it is difficult for cyber insurance providers to effectively determine the pricing for a policy. Many cyber liability insurance companies utilize security companies to perform security assessments. Unfortunately, all but the most astute individuals will be unable to effectively judge the risk associated with a particular environment unless the vulnerabilities are blatant. Many security specialists simply look for actively practiced best security practices, without delving into many of the issues identified above or try to provide high level assessments which have limited value in practice, exposing only the most basic flaws. They often assess surface security issues rather than the commitment and ability of an organization to evolve its security capabilities on a continuous basis. Without the solid commitment to security, even the most secure organization today can lapse into security mediocrity within a few weeks of normal Information Technology operation.

To this point, many of the Internet access points (Internet Service Providers and the like) have not had to shoulder the burden of being liable for illicit activities conducted through their networks. In some countries this is law, in some it is simply practice. Since there is no reason to assume the liability, these types of companies do not often put in place the necessary security technologies, measures, and training to keep their environment secure. Even if we successfully enact these types of laws in the United States, there may be other countries that are unwilling to comply. Since this is a worldwide series of networks, it can be hard to enforce United States laws and standards on other countries, especially those opposed to the United States. Even recent attempts to pass cyber crime laws in the Ukraine and Belarus have failed. In short, we can not expect all others to shoulder the liability for protecting their networks from malicious activity.

The Spoils of War

It would be fallacious to state that there is only a single reward that awaits those entities that are successful at waging this war. In fact, there are many potential rewards but virtually all of the rewards are on the side of the entity that acquires information, whereas few rewards exist for those trying to defend their information resources. As we discussed earlier, there are many individuals that do not like the United States, many individuals that are bent on committing crimes, many individuals who are seeking to assert their political or religious agendas on others, and many individuals bent on harming others. Although this is not a purely American phenomenon, America does present an appealing target. Why? It is because of our wealth and our propensity to rely heavily on automation and Information Technology. Few people in the world live as we do in the United States. We utilize a disproportionate share of the world's resources and far too often we live in wanton luxury, especially when compared to many people around the world who do not have access to even the basic necessities such as food or health care. Many people throughout the world would like to see the United States falter, thus we become an appealing target. We have also involved ourselves in affairs in areas of the world where we have considerable religious, political, and philosophical differences raising our visibility as a target.

Democratic entities such as the United States have traditionally been strong internally and weaker externally, thus, attacking our economic infrastructure has been very successful over the years. Although many people died on September 11, 2001, fallout from the attacks continues to damage the economy. Individuals within our country now know that we are no longer immune to terrorism on a grand scale, even within our own borders. The result? Many individuals have halted their purchasing patterns. Many have stopped investing in the stock market even though interest rates are historically low. Many have delayed even needed purchases until more confidence in the economy is restored. This should be a wake-up call for us all.

The rewards of successful Internet hacking are many. First and foremost are the financial rewards that can be obtained from compromising information that flows across the Internet. Credit card numbers are routinely stolen and have resulted in an increase in credit card fraud. Many consumers share the opinion that using limited liability cards protects them from losses. Nothing could be further from the truth. When credit card companies lose money, they don't just accept it; they raise their rates to the retailers or increase the costs of other services to compensate. The retailers, having to now pay more for their ability to accept credit cards, raise the price of their goods. And then of course, the consumer pays more for the products they purchase. Whether we like it or not, whether it is visible for all to see, we all pay for credit card fraud in one way or another even though it may not actually show up on our credit cards bills as waiving unauthorized charges incurred by others.

The <u>2002 Computer Crime and Security Survey</u> published by the FBI reported that 223 companies who were able to quantify their losses estimated that they lost more than $455,848,000 and that 74% of the attacks launched against their systems and information were launched through their Internet connection. While this number is alarming, it is not a true representation of the damage done for two reasons. First, collateral damage often results from an attack that does direct damage. If a company loses money through information being stolen, the same attack could harvest individual customer information which may be used at a later date. This is part of the collateral damage spoken of. Second, many companies do not report their attacks, especially if they involve known vulnerabilities. Thus, the report covers those companies who have sought assistance in solving their cyber crimes.

Information access is another big payoff. There are certain types of information that flow across the network that can be used to wreak havoc. Access to information that affects national security, reveals the locations of military bases, depicts nuclear plant locations, defines security policies and security processes, describes war plans, or compromises United States activities only helps our enemies. They will use information that we allow them to access, against us. During the cold war Nikita Khrushchev once said "We will hang you...and you will sell us the rope". While we don't necessarily sell our adversaries information that will hurt us, (exception: the well known social engineering failures that allowed sensitive information to fall into the hands of our enemies) we also don't always protect it adequately. This information can be used against us in a number of ways. Information, especially sensitive information, must be protected regardless of whether or not it is sensitive information associated with a consumer, an employee, a company, or a government.

Once a hostile entity has access to information, they may be able to modify the information. Depending on the nature of the information accessed, considerably more harm could be done by modifying information as opposed to stealing or destroying the information. This can be especially valuable if the owner of the information relies on the information to transact business, wage a war, or manage resources. Just consider for a moment the impact of key information that might be modified. Medical information, account balances, credit scores, and criminal records could all have a devastating impact on an individual if they were modified in a negative manner.

Another spoil of this type of war is embarrassment. We tend to think of ourselves as technologically advanced; at times we take it to the point of arrogance. An article in the October 8th, 2001 Guardian Unlimited web-based news paper addresses this succinctly, claiming that the assumption of the West's (as in western society) superiority is

179

as dangerous as any other form of fundamentalism. We must guard against hubris so that we do not fall victim to overlooking or underestimating our enemies. We can easily rationalize that our technology is the best in the world and thus, we have nothing to worry about. We must guard against this at all costs and accurately assess the capabilities of our adversaries. Although God has blessed this country greatly, we must never take for granted what we have. Vigilance is required at all times. When we view ourselves as invulnerable because of our technology, we open up the door to be embarrassed. Out enemies want to see us stumble, they want to see us fall. When the attacks of September 11th occurred, we were angry, we were scared, and we were embarrassed. The security that we had traditionally enjoyed in America was gone. Our technology, our processes, and our geographical safety net all failed to protect us against that particular set of incidents.

A War Capable and Worth Winning

Clearly, the Internet can be used to share information, but as we have shown in this book, it is often not suitable for the exchange of sensitive information. As long as the Internet remains a network where the population is unknown we will never know who is looking over our virtual shoulders, who is seeking to break into the systems attached to it, who is devising the next mega-virus, and who is seeking to use our own information against us. We, as a society, must simply rethink our value system when it comes to securing our most important asset – information. Whereas many of the other values and constructs in our society are open and easily accessible, security should not be one of those openly discussed, promoted, or implemented items. We need to acknowledge our participation in the war and transform the existing war by changing the way in which we wage it.

Each company should consider whether or not securing resources is within their core competency. Some organizations might argue that this is a necessary part of any business and perhaps rightly so, but if a company determines that security is going to be a part of their core competency, they should invest in that core competency so that they are able to execute upon this competency decisively and effectively. Companies should be well advised though: this investment is significant. If a company determines that trying to secure their informational resources is not within their core competency, other options exist to try to ensure these resources are well protected.

There are many ways that we can establish the ability to wage the war in a manner that we have a much better chance of success in the war. There is no single activity that we can undertake to guarantee success in the war; it takes a series of activities, carefully thought out, to establish a war scenario which improves our chances of waging it successfully. It will be necessary to radically alter our thinking about some of our most taken for granted ideas and concepts. It will take leadership that is willing to step out from the norm and establish a more effective way to handle security. And, it will take time. Let's look at some of the changes that we need to consider enabling us to improve our chances of waging the war at least partially successfully.

A New Battlefield

Since we have seen throughout this book that it is going to be next to impossible to compete with the active and hostile elements that exist on the Internet (at least with our existing levels of staffing) we either need to change the battlefield or invest much more heavily in our security resources. If we are going to try to do this in a cost effective manner, the war needs to be fought under conditions so we have a chance to win the war. How? By moving the war to a battlefield where we have a chance – to a private network infrastructure where the population of the network is a known entity and where only valid business interests are engaged. On this type of battlefield, we or the providers of such a network infrastructure can exercise much more control over what flows over the network and who can have access to the network. As discussed earlier in the book, just as we have in our application architectures moved back to very limited function desktops (browsers) so it makes sense to consider returning to the use of securely managed private networks, but not in the spaghetti styled manner that we did in previous generations.

Reducing the number of threats is one of the key approaches in trying to secure anything. Think about the choices made by NORAD (North American Aerospace Defense). Why did NORAD choose Cheyenne Mountain as one of its command and control locations? It is because they could reduce the number of threats against the facility, thereby improving the overall security of the facility. The facility is impervious to many modern day weapons, can not easily be compromised physically because of its depth under solid rock, and even has its own set of utilities in the event that external utilities are compromised. That does not mean that it is invulnerable, just that they have taken steps to reduce the threats to manageable levels.

It is considerably more cost effective and more secure to utilize private networks for the exchange of sensitive information than continually fight the war to protect sensitive information on an open network, achieving varying degrees of success. Exchanging information over a private network, at least a well managed private network with an effective security management staff, will allow us to exchange sensitive information in an environment where the members of the network community are known. No longer are we dealing with an unknown population that may be hostile in nature. Private networks allow us to minimize the number of threats that exist against our information resources because virtually all hostile elements would be weeded out of the private networks and because as we have seen earlier in this book, it is access that allows hostile elements to attack our resources.

There are ways to set up private networks that are more cost effective than the Internet, but are much more secure. These private networks would ideally be managed by a group with extensive experience in network and systems security, social engineering, and would use an unpublicized set of security protocols. The networks would be monitored on a continual basis ensuring that aberrant behavior would be flagged and that traffic terminated. This centralized entity would know the constituents connected to the network and systems of its constituent members and could manage change control for these systems so that before any significant technology is introduced into the environment, they could be thoroughly tested to determine any impact to the community at large. The same would happen before new applications are deployed.

Using security companies to help us develop security models within our own companies is of value, but another potential solution is to rely on SSPs (Secure Service Providers) to handle the exchange of sensitive information. An SSP provides a much more secure landscape than exists in an open, public network. In addition, a SSP can provide a myriad of additional services that are not generally available between different entities exchanging sensitive information. SSPs execute their core competencies in a diligent manner since their company reputation depends on it. They attract highly skilled individuals who improve the overall chance of keeping sensitive information exchanges secure. Using SSPs which operate over private network infrastructures greatly reduces the exposure to attack since the landscape over which information is exchanged is not available to all.

The number of highly talented, experienced security individuals within the industry is somewhat limited and those entities that are fortunate enough to have them are fortunate. Unfortunately, many organizations do not enjoy this luxury. As companies try to adjust to the ever evolving threats over an open network, even more qualified security specialists will be needed. Over time, this may turn out to be one of the most severe shortages in terms of a talent pool that has ever existed, and one of the most critical. SSPs would likely attract highly talented and highly motivated individuals who are interested in security, thus they facilitate development of critical masses of security experience. The table below contains estimates for the number of

185

security personnel required to accomplish key security tasks inside of an average organization.

Security Tasks	1 Company	100 Companies	SSP
Monitor New Vulnerabilities	0.50 FTE	50.00 FTE	5.00 FTE
Determine Impact of New Vulnerabilities	0.50 FTE	50.00 FTE	5.00 FTE
Develop Plans Against New Vulnerabilities	1.00 FTE	100.00 FTE	10.00 FTE
Test Applications for Security Vulnerabilities (depends on development staff, but scales)	1.00 FTE	100.00 FTE	100.00 FTE
Test Operating Systems and Infrastructure Software for Security Vulnerabilities	1.00 FTE	100.00 FTE	5.00 FTE
Introduce and Configure New Technologies Into Testing Lab	0.50 FTE	50.00 FTE	5.00 FTE
Test New Technologies in Testing Lab	0.50 FTE	50.00 FTE	5.00 FTE
Provide Feedback to Vendor	0.10 FTE	10.00 FTE	1.00 FTE
Setup and Configure Network Security and Monitoring Devices*	0.25 FTE	25.00 FTE	25.00 FTE
Setup and Configure Security on Data Sources*	0.25 FTE	25.00 FTE	25.00 FTE
Setup and Configure System Level Security*	0.25 FTE	25.00 FTE	25.00 FTE
Review System Logs	0.20 FTE	20.00 FTE	10.00 FTE
Review Application Logs	0.10 FTE	10.00 FTE	10.00 FTE
Review Physical Access Patterns	0.10 FTE	10.00 FTE	10.00 FTE
Perform Code Reviews	0.25 FTE	25.00 FTE	25.00 FTE
Review Traffic Patterns, Looking for Attacks	0.50 FTE	50.00 FTE	5.00 FTE
Develop Security Policies	1.00 FTE	100.00 FTE	5.00

Security Tasks	1 Company	100 Companies	FTE SSP
Provide Education on* Policies and Social Engineering	0.25 FTE	25.00 FTE	25.00 FTE
Develop Social Engineering Strategies and Implement	1.00 FTE	100.00 FTE	5.00 FTE
Total Number of FTE's	9.25 FTE	925.00 FTE	306.00 FTE

* Somewhat variable, depending on number of devices.

Although the actual number of people performing security activities will depend on the size of the company and the nature of its business presence on the Internet, it is clear that there are definite economies of scale by consolidating these functions.

When you look at the chart, you notice there is not necessarily a linear savings for all functions. Why? The reason is that certain types of activities can not be scaled down across companies. An example of this is the task *Test Applications for Security Vulnerabilities.* If the delivery rate of applications to a security testing entity requires 1.0 FTE, there will be no savings across 100 companies since the workload compounds because the number of applications to be tested also grows linearly (assuming development staffs in each organization). Alternatively, the task *Test New Technologies in Testing Lab* can be reduced dramatically because there is considerable overlap between the technologies used across the 100 companies.

Look at the dramatic savings in security staff by centralizing these functions. Let's face it: we have the same people, doing the same jobs, in different companies. When we don't have the same types of people doing the same jobs, it is often because we aren't doing that function at all. A centralized entity managing a secure network infrastructure would attract a considerable number of top level security professionals. This type of approach is not designed to render the Internet useless, but the Internet should be relegated to less important and less sensitive traffic than is currently being sent over it.

One of the other considerations that will always arise is the consumer. Consumers generally attach to the Internet and so many companies feel that they must also connect to the Internet to reach their customers. Unfortunately consumers are also under attack and the Internet is not a particularly safe place for consumers to roam either, although their systems are often of less value than corporate systems and present a reduced value target. Within a corporate information repository an enemy may be able to acquire 1,000,000 or more credit card numbers within a single attack. Hacking into individual systems still requires some work but with a considerably reduced payoff: perhaps one or two credit card numbers may be acquired. Other information from a personal system could also be acquired though. In addition, consumers systems can be hacked into for other reasons. Once an enemy gets into an individual's system, they can then use that system as a platform from which to launch an attack without the knowledge of the end user or it can be used as an intermediary routing system to a machine that will actually launch the attack.

There are ways to handle these issues as well. We might consider the use of a consumer based network, where the population of the network was also a known quantity, where authentication was required, and which could be used to conduct business between individuals and business. While it would take time to develop such a network and populate the network with both users and content, it could be done in an evolutionary manner allowing the existing Internet to continue to operate, but eventually we would be in a much better position to secure all of our resources. Government transmissions of sensitive information should never traverse the Internet. A government services network, separate even from the corporate and consumer networks should also be considered.

Consider the reason that hostile elements run amok on the Internet. It is primarily because of their desire to get access to sensitive information which they can use in the commission of crimes. If we slowly begin to remove the sensitive information from the Internet or from devices indirectly connected to the Internet, the Internet slowly loses value as a target.

There is another way the battlefield can be shifted. Individual companies often use a wide variety of technologies to support their business processes. Some technologies have better security records than others. Companies have the option to select what technologies they use. They can choose more secure or less secure technologies. Often, technology selection is a political process, overlooking the primary need for security. Companies must seriously re-evaluate their usage of technology with a primary driver being security. This is what they will eventually be judged on.

When companies use a large number of different technologies, this complicates the security scenario. The vulnerabilities that exist in one product are often different than the vulnerabilities that exist in other products. This is a classic example of a double edged sword which can cut both ways. If we use a single set of technologies that has many different vulnerabilities, we have a problem in trying to keep those vulnerabilities closed. If we use multiple sets of technologies (e.g. from different vendors) there may be the opportunity that some of the strengths and vulnerabilities in each set of technology may offset one another, but unfortunately the more different technologies you use in your environment, the more differentiated skills sets you need to maintain in order to try to keep your environment secure.

Moving from Reactive to Proactive

We have seen throughout this book that it is very costly to try to be vigilant and responsive enough to deal with the array of different attacks launched by well funded and well organized entities. We have seen that even if we were able to adequately respond in a timely manner, we still have the initial exposure that is created when a new attack type is launched and the window of vulnerability we have from the time an attack is identified to the time when all vendors and customers have implemented protective measures against the attack. With the advancing speed of viruses moving through the Internet, the cost of being reactive is going up.

It would be an interesting study to understand how many organizations had taken the time to develop a proactive approach to security by trying to list all of the potential exposures and then developing proactive countermeasures for each of the exposures. I believe if this activity were undertaken in corporations, many executives would begin to understand what they are up against. Effective security begins at the top where executive leadership has established a "security first" mentality among company members and which allocates the necessary resources to protect a company's information resources. This is the first step in assuming a proactive security posture.

Large organizations can be notoriously hard to reorient with respect to security and yet the larger the organization, the more vulnerabilities exist and the more the need to focus on security. Larger organizations often have more people to exploit, more ingresses to enter through, more systems to compromise, more changes to the infrastructure through which vulnerabilities can be introduced, and most importantly, more sensitive data, thus larger organizations generally present a higher target value.

The migration from a reactive security posture to a proactive security posture is neither economical nor trivial. Reactive security measures are those measures that are executed in response to a threat, whether real or perceived. Reactive security postures are event driven and leave us vulnerable to the initial set of attacks just as a counter puncher is susceptible to the first punch if it is landed by an opponent. Viruses and the like can be much more invasive though. Eventually, the first strike by a sufficiently advanced attack could render an organization significantly impaired for a considerable period of time. If the necessary business continuance protocols are not in place within the organization, the time that an organization is impaired could be even greater, with the potential that data could be permanently lost.

Although almost all organizations implement some level of proactive security such as an Intrusion Detection Device which is designed to detect and terminate attacks before they are successful, there continue to be many aspects of our security which are still reactive. Examples of these reactive security measures include anti-virus software and the review of historical system and network activity logs. We utilize these types of security postures and techniques largely because of limitations that exist in the technologies that we deploy and our lack of training and process.

Proactive security measures on the other hand are those designed to prevent an event or occurrence before it even begins. While it will not be possible to fully envision and counter all threats, many security activities can be handled in a proactive manner. In order to begin our journey towards proactive security, we have to address a number of different areas. In addition, even after we make the initial set of changes in our organizations, continuing investments will be required to maintain the proactive security posture.

The first of these investments is in training. The goals of many of the proactive processes that we discuss are to allow us to discover and correct flaws before others can find and exploit them. This positions us to be able to remove many avenues of attack before they can be exploited, although as we have stated, it will never be possible to fully discover and close all vulnerabilities. Security specialists must be trained in a wide array of technology and process disciplines to effectively review, analyze, and defend against different types of threats. The training must be continuous since knowledge in the security world can be dated very quickly. Security staffs need to be trained in almost all technologies used on a regular basis by the organization. If the sphere of knowledge that a security staff member receives is limited, their vision into what really goes on is impaired as is their ability to secure an organization's resources. Good security staff members, as we discussed earlier, have both breadth and depth of knowledge and experience.

Training investments also need to be made for developers and individuals who deploy applications. Care must be taken to ensure that we develop secure applications and that those individuals that perform release and configuration management within our organizations understand the implications of the applications they are deploying. Once again, both of these rely on training and understanding the fundamental security issues around developing and deploying applications.

For a developer in Information Technology, a required curriculum covers such items as application security strategies, secure application designs, security architectures, secure usage of network services, validation techniques, authentication, non-repudiation, authorization, secure application coding, notable application vulnerabilities, etc. In addition, the curriculum should cover the nature of sensitive data, what is classified as sensitive data within an organization, preventing the transmission of data in clear text, and implementing security standards within their applications.

This is one of the greatest fundamental flaws that exist within the current environment. We continue to churn out software at an amazing rate, but the real question is: how much of that software is really secure. Training the people developing and deploying software is one of the most important things that we can do. Each day that passes brings us new, insecure software which makes its way to the marketplace. This ever increasing amount of software will either have to be scrapped or fixed at a later point in time. While it is not possible to stop all development, now is the time to invest in the training necessary to help all of our developers create secure software so that we are quickly evolving towards the development of more secure software.

Training is also required for systems administrators, database administrators, network engineers, and other operational personnel such as the help desk and end user support personnel. Curriculums for these types of individuals generally center around the technologies that they manage or support, but an awareness of general security principles is required for these types of individuals as well.

Individuals that are not a part of the Information Technology organization also require training. For individuals that don't reside in a security or Information Technology function, it may simply be a education of social engineering issues and an awareness that extends to items such as user ID's and passwords, the guarding of sensitive information, and the protection of corporate resources and information. It is much less costly, and provides much better protection, if individuals within an organization know beforehand what information they can and can not release, under what circumstances it can be released, and to whom the information can be released.

Once the initial round of training has been completed, refresher courses are needed on a regular basis to provide updated information to personnel. This is one of the most important and proactive things that an organization can do – get their personnel in a position that they not only understand and keep current on security, but that they are trained well enough to be able to effectively participate in the overall security of the enterprise. Just as with a chain, the security of an organization is only as strong as the weakest part of its security infrastructure.

The second of these is investments in process. There are many processes that can protect us. Processes that need to be considered and which have direct bearings on security include change control processes, testing processes, security design patterns for applications development, code reviews processes, social engineering protocols, technology evaluation processes, and so forth. Each of these processes should be implemented in a methodical and proactive manner. While a sufficient level of process can be a hard thing to determine, sneaking up on the sufficient level of process is risky since probing may be able to discover weaknesses in process. In addition, there can be many failures while taking the time to slowly more towards the proper level of process.

Development of the change control process should be implemented within a company and also with business partners that share the technology undergoing the change. In addition, not only should changes which have the potential to impact the security of an organization's resources be put through a controlled process, they should also be tested prior to the implementation of the change. We can tend to underestimate the impact of small changes on the security of the environment. In many organizations, the security staff may not even have the chance to review proposed changes in key items such as firewall configurations, database security schemes, etc.

A formalized set of testing should occur against all developed applications using an environment that is a mirror of the production environment to which the application will be deployed. The testing should check for standardized application attack strategies such as buffer overflows, but also should check for more insidious attack strategies validating that inappropriate logons and logins or back doors are not left in the code. It should also test for the impact of any required configuration changes in infrastructure components such as web servers, database management systems, etc. In addition, a regression test suite should be developed ensuring that all test cases related to security exist within the regression test suite even if only a percentage of the functional requirements exist in the regression test suite. Once the initial security regression test suites are developed, they need to be extended over time though the results of internal testing of technologies and applications as well as vulnerabilities identified from external sources.

Design patterns are a method of being able to communicate experience and best practices between development team members. Focus should be placed on developing security design patterns and *secure* functional design patterns for developers of software. Even when these are created though, general distribution of them exposes us to the risk that our adversaries get their hands on them, assimilate them, and find a way to exploit them. These design patterns can take the form of specific patterns that are defined for input and output, authentication, validation, encryption, data partitioning, etc. Integration design patterns also need to be developed to focus on the exchange of information between different systems in a secure manner and defining how credentials are passed between systems as well as how to limit access to the minimal amount of required data. There are many different avenues to attack applications so creating and updating an applications design pattern and best practice document and code can provide significant value.

Code reviews should be implemented to expose developed code to a wider set of eyes. This can prevent such items as security back doors, Trojan horses and other code based violations of best practices. Thus, not only can bad coding practices be caught and dealt with before deployment, but the process ensures that no malicious code is intentionally or inadvertently embedded within a developed application. A company should view suspiciously, any developer who is unwilling to submit his or her code to closer scrutiny by a larger audience of developers. While this does not guarantee prevention of malicious code being introduced into software, it does reduce the chances of this succeeding since the compromised code would have to be kept secret amongst a much larger group of individuals. This is one place where group dynamics work in the favor of the organization. It is difficult to keep things hidden within a large group.

A technology review process is also a prudent process to enact. The goal of the technology review process is to research and test new technologies, specifically from a security focus, before they are implemented in a production environment. While this often calls for multiple lab environments, the investment in the process can be well worth the time. We still encounter the problems that we discussed earlier, namely not being able to envision all potential vulnerabilities, but at least this gives us a chance to test known vulnerabilities and to potentially discover new vulnerabilities before they are exposed to an open community. If the security staff has sufficient experience, they can consider and develop new potential attacks in order to test the technology, perhaps using some of the same tools that our adversaries would use against us.

This covers only a subset of the necessary processes that will be required to deal with security related issues. We have not even touched upon the various social engineering processes that need to be developed, nor have we begun to address the processes needed to deal with fringe attacks. As security technology gets more robust, there will be a corresponding increase in the number and complexity of the social engineering attacks. Remember that most technologies can easily be defeated by the appropriate social engineering attacks. It is for this very reason that we do not want to wax eloquently on a litany of potential social engineering attacks.

Third, and finally, we need to take a more proactive approach to technology selection and implementation. We often select technologies without considering security as a prime consideration. There are technologies on the market today that have the ability to reduce our vulnerabilities in multiple areas. A simple example of this might be a web only desktop which doesn't run a traditional operating system and which relies on servers for access to popular applications. It is almost always easier to secure back room servers rather than trying to secure desktops which often come complete with removable media through which hostile code can be introduced.

This brings us to another of our security axioms: **The ever evolving nature of the war requires a proactive mentality, attempting to identify and close all possible vulnerabilities before applications and technologies are deployed.** We won't be totally successful at this type of approach, but we must make the effort to identify and close vulnerabilities before our enemies exploit them.

Listed below is a list of proactive activities which can be performed within an organization to adopt a more proactive security posture.

Performing these activities does not eliminate the need for reactive activities such as historical log review, which still must be undertaken, but it does position an organization to keep as many of the potential vulnerabilities out of the public arena as possible. This list is designed for those who have intentionally determined that they want to continue to operate on an open and unsecured network. Other solutions will be discussed later which can mitigate the need for some of these measures. There will always be other things that can be done, but this list forms the basis of a good start.

List of Proactive Security Measures
Create a test environment and use it to test all changes scheduled for implementation in the production environment including new applications, application deployment techniques, software maintenance, new technologies, and configuration changes.
Require positive identification scheme for all security related transactions within an organization (no exceptions)
Develop a best security practices document covering the development of applications and update the document with new vulnerabilities as they are found, then use it to develop future applications. Cover all security related items such as database access and storage, user authentication, user authorization, communication between components, state management, session management, etc. Have a security design plan and review, accompany all new development.
Develop a security regression test suite, augmenting it as new vulnerabilities are found and execute the test suite against every application augmenting it as new vulnerabilities are found by both internal and external entities
Obscure as much information flow from the user as possible in web based applications; maintain no state in components such as browsers or web servers
Utilize qualified external companies to try to break your applications in as many ways as possible; don't just limit them to penetrating the perimeter for the sake of cost
Establish technologies, especially at Internet ingresses and egresses that monitor traffic and can take action at the first sign of potential problems.

Staff your security team sufficiently to handle the tasks necessary to protect the enterprise in a proactive manner; empower them to implement inconvenient practices consistent with best security practices. Invest in continuous training of your staff insuring they are kept up to date with changes in the security world.
List of Proactive Security Measures
Configure every device (systems, network nodes, etc.) with the minimum functional complement of access (e.g. open the smallest number of holes in a firewall that will allow applications to work). Audit these devices on a regular basis for outdated or obsolete access points
Keep all systems and network components at the most current level of maintenance closing known vulnerabilities

Proactive security doesn't stop with the list above though. There are a large number of additional measures that need to be taken to adequately protect the resources of an organization in a proactive manner. Our goal is to beat the hackers and criminal elements to the vulnerabilities, effectively limiting the damage that they can do. If we beat them to the identifying, exposing and closing the vulnerabilities ourselves, we position ourselves a step ahead of many of their efforts. Even though we can't catch them all, each one that we remove improves our overall security posture.

Limiting the Damage from Attacks

Consider for a moment critical infrastructures that have developed naturally within this country over time. Examples of critical infrastructures include the highway system, the power distribution system, the water supply system, the banking system, etc. These critical infrastructures were developed over time in a very diversified and compartmentalized manner. While this evolution occurred naturally, this structure provides considerable insulation against attack and inherently limits the amount of damage that can be done by a successful attack to the overall critical infrastructure.

There are very few (and none in some critical infrastructures) locations that a terrorist or other criminal element can physically attack to disable or eliminate an entire critical infrastructure. If a terrorist blew up a power plant (even a nuclear power plant), damage would be limited to a geographical area and a limited number of individuals would probably be without power (for this discussion, we will ignore the potential casualties from the attack). Consider this from a different perspective. Imagine that the United States had decided to build a single centralized water supply to meet the fresh water needs of the nation. While this may be impractical, it does illustrate a point. If there were a single national water supply this would present a very attractive target to a terrorist since they could affect such a large number of people by breaching a single piece of critical infrastructure. In addition, since all water is controlled through one distribution system, it is no longer necessary to spread out resources trying to breach a large number of systems that are secured through different methods with the potential to cause a limited amount of damage to each. The same is true as relates to information. This brings to one of our security axioms: **Effective network and information partitioning strategies inhibit attacks by reducing the target value and reducing the damage from successful attacks.**

This is of considerable interest on the Internet. Although many would point out that the Internet is very compartmentalized in that it is made up of a large number of different physical networks, the reality is that the Internet is not really compartmentalized from a logical perspective. In reality, there are many holes (TCP ports for example) that are open between different physical networks and it is these openings that various attacks use to move between those physical networks in a logical manner. For example, most corporations have port 80 and port 443 open between the Internet and the DMZ (DeMilitarized Zone) and many organizations have ports 80 and 443 open between their DMZ (DeMilitarized Zone) and their internal networks.

Partitioning or compartmentalizing information and resources inhibits attacks because each individual attack can only affect a portion of the total, thus it is a lesser target. If we carried this concept through to an extreme point, by partitioning a 100 character medical record into its constituent characters with a key and storing them in 100 different databases each with different security models, how many hostile elements would have the patience and fortitude to try to break into the majority of these systems, just to have a reasonable chance to assemble some meaningful information? In addition, it reduces damage because if a portion of the set of information is compromised, damage is done only to a fraction of the total data; the rest may remain safe. In this example even if 20% of the individual databases were compromised, it is unclear whether any value would be obtained from the break-ins. If the 20 characters covered the last name and the medical condition, value might exist, otherwise most other combinations are likely valueless. A hacker would also be required to reassemble the constituent parts of the data for it to make sense. If they only got access to 20 random elements, it might be hard to determine what that information is.

There are a number of different ways to limit damage from an attack whether it originates from an internal network node or an external network node. First and foremost, an internal network should never be a trusted network where authentication in one set of systems automatically grants access to other systems. This is important for external attacks as well since an external entity able to compromise the external security and obtaining access to one system will be able to get to other systems through these trusted relationships, expanding the amount of damage that can be done.

Secondly, damage from an attack can be limited by partitioning. If an attack occurs, this can limit the amount of damage that can be done to the environment as a whole. Many organizations tend to lump together many different data types into a single database and then utilize a common security model for all of the databases. Imagine for a moment that all information worldwide was stored in one database. Although this also clearly impractical, it does illustrate a point. Once the frontline security for that database was compromised, the intruder might have access to all information in the database and would be able to cause significant harm. While this principal can be taken to extremes by separating all information in different databases, a more reasonable and sound strategy is to store information that must be used in tandem, separately.

An example of this would be to store medical histories in one database with a common key and people in a different database with the same key but with separate security models and credentials required for each database. If the database with the medical histories is compromised, there is no knowledge of who has what condition. If the database with the people is compromised, only information on the people would be available. There would be no connection to their medical histories unless both databases were compromised. One of the places this is often violated is in customer databases. Far too often, financial information, and the associated personal information are stored in the same databases so that when one is compromised, the criminal elements have enough information to illegally obtain goods and services through credit card fraud, funds transfer, etc.

A solid security best practice is to avoid our natural inclination to combine things into larger things, especially for convenience's sake. This only exposes us to greater risk. We have accepted the Internet, not because it is the best place to do business, but because we value universal connectivity of greater worth than secure electronic commerce. Whenever we attempt to centralize information or resources, we make our systems higher priority targets and we increase the overall damage that can be done to the resource if it does get compromised.

A carefully thought out division of labor strategy needs to be implemented for internal systems. This division of labor should be multidimensional in nature, separating access to logging information from systems and network administrators. Systems administrators should not have access to network devices and network engineers should not have access to systems. While this is simplistic, it does depict the nature of the division of responsibilities. Separating out the logging responsibilities on both network and systems should ensure that an attack from a single individual can be tracked. Partitioning other administrative responsibilities limits the amount of damage that a single individual can do. Division of labor is one of the key social engineering activities that should be addressed in any organization. No single employee within an organization should be in the position that they are able to circumvent an organizations security models with only the knowledge associated with their job. If one person can change fundamental pieces of a system or network and then wipe out any traces of the changes, the responsibilities have not been partitioned across enough individuals.

How does a social engineering protocol of this type get developed? The social engineering protocols gets developed by functionally breaking down the administrative and security functions so that the security staff can not make administrative changes and the administrative staff can not make security changes. Clearly, there is much more work in the social engineering area than this simplistic example depicts, but it serves to demonstrate a level of effort that many organizations have not yet undertaken to protect themselves from internal threats.

Obfuscation

One of the most obvious ways to secure things is to obscure them. Obscuring things is an accepted way to protect something. The Internet is risky because the common theme of access is available. We can easily obscure the nature of information flowing over the Internet, but not easily hide the transmission of information. We, as a culture, have become so enamored of openness, we apply it to areas that it has no business in. We provide architectural layouts of facilities and criminal elements use these to enter and exit facilities in the course of committing crimes. We report on how a person was able to violate security at an airport gate and others make similar attempts or develop a hybrid strategy which may not be caught with the same security mechanisms that caught the initial person. We talk openly about how a given scam works and then others develop variants of the scams to entrap others. We publish cyber security standards for all to comment on, exposing our security strategies. We publish a list of vulnerabilities and how to exploit them and criminal elements add these to their arsenal, catching all but the most diligent entities. We publish code that hackers can use to exploit vulnerabilities and we are surprised at how effective their attacks are. We communicate and store sensitive information on an open network.

Rather than distributing security information to everyone, we should distribute this information carefully, allowing access only to authenticated individuals. This also goes for security models, the exposing of vulnerabilities, and social engineering activities. An example of the correct process for handling a discovered vulnerability in a proactive manner was the security hole recently found in Sendmail. The vulnerability could allow an individual to take over a Sendmail server allowing the launch of Denial of Service attacks through a single email. Initial profiling suggested that more than 50% of the Internet mail servers could have been affected. However, unlike the normal notification process, the hole was actually discovered in December of 2002, but the posting of the vulnerability was delayed until early March 2003. This provided some time that key users could be notified of the problem and work could begin on mitigating the risk. Thus, the clock did not start for hackers until March 2003 when the vulnerability was announced hopefully reducing the window of vulnerability. The hole, which was found by Internet Security Systems, was shared quietly with the Government who began to coordinate mitigating strategies. This is exactly the approach that is required in dealing with vulnerabilities.

We are fortunate that this is a vulnerability that friendly forces found and acted on appropriately, otherwise we could have had a major problem. Exposing vulnerabilities and defining security protocols in the open only exposes us to those who would harm us and sets off a reactive set of steps to try to fix the problem. By not releasing the information to all, the opportunity to address the vulnerability in a proactive manner was at least possible. Information of this nature should be released only to authenticated individuals, not as general purpose information that anyone can get access to, regardless of their intent.

Transmission of sensitive information has no place on an open network, especially one teeming with hostile elements. While an open network does provide some advantages, the risk and the potential liabilities of using an open network are completely outweighed by the disadvantages. Enemies of this view might argue that encryption, in effect, obscures transmission of sensitive data which is true, but it also acts as a beacon, calling attention to the encrypted packets. With so many of the packets traversing the Internet being inspected, those that are encrypted are given special attention as they are assumed to contain sensitive information. While encryption does mask the nature of what is contained in transmissions, people rely on it too heavily and assume that it completely protects the data, which is not necessarily the case. So while encryption does help to hide data from plain view, a much safer course of action is to not exchange the information in the open at all, but to exchange information in a secure or closed environment.

In addition to encrypted transmissions, there are some products that transmit information in clear text. Examples of this type of product are some instant messengers. Many people do not realize that many of the Internet based instant messaging products transmit information without encrypting it. When it is used in a business environment where the instant messaging server is not wholly located within the firewall of the business, information flows out to the Internet and then back into the corporate network. People often assume that if the conversation is internal to the company (e.g. between company employees) that it stays within the corporate network. I have seen organizations that exchange sensitive information such as passwords over instant messaging products using the Internet-based servers of the company providing the instant messaging product. This is clearly no way to exchange sensitive information.

Another risk from the openness perspective is the use of standardized security protocols. In reality, it is arrogant to assume that any individual or group of individuals have the ability to devise a foolproof security model. As we discussed above, the ability for an individual or a group of individuals to consider every potential threat and thwart those threats through technology or social engineering is simply not realistic. For too long we have tried to utilize the Internet for everything instead of making use of it in a judicious manner. It would be very similar to trying to fix everything with only a hammer. We do so because it we can connect to everybody with one connection, a very convenient network indeed.

Mentally, we have to make the shift to be able to at least consider utilizing different networks for different applications. We simply must begin the process of transitioning sensitive data transmissions to a series of networks that are not generally accessible and where the population of the network is known. Hiding information through encryption can be effective, but doesn't give us the best chance to protect information and with so much at risk, we need to select strategies that give us the best chance to protect our critical information resources. This brings us to the last of our security axioms: **Obscuring information and information transmission can be an effective way to protect information.**

Security through obfuscation is good, but this can and will work against us as well. Obfuscation can also be used by our enemies to hide the nature of their attacks. In most cases, the tools that are available to us are available to our enemies, even though there are technically restrictions on exporting some technology. When they acquire security technologies, they can begin work on reverse engineering the product (this fits for both hardware and software) to see how the protection device functions and to determine how to bypass the security it affords (remember the first step in defeating a security scheme is to know of its existence and its nature, therefore obscuring a security scheme provides some inherent protection). When they get their hands on the very same technology we use to protect our information and resources, they take the first step in this process. They can also test new attacks in the safety of their labs without exposing themselves to persecution or capture, releasing an attack only when it is fully evolved and ready for release. Catching someone in the process of developing malicious code is very, very rare unless they are being pursued for some other attack they had previously launched.

When our enemies use obfuscation to launch their attacks, they do it to protect the location of the launch point, to fool our protection devices, and to maintain their anonymity. They will continue to use these tactics against us. Our enemies also use our well known protocols (e.g. IRC and HTTP) to be able to send information to hosts that have been compromised. They will continue to use obfuscation as a method to intersperse attacks with legitimate traffic so as to be able to prevent us and our tools from telling the difference.

I believe this is one of the attributes of the next generation of attacks – taking hostile packets bound for our networks and making them look legitimate. As we continue to try to develop predictive software and anticipate what our adversaries are going to do, we are going to be increasing faced with false positives. False positives are attacks that are detected against our infrastructure by certain technologies, which in reality are legitimate business transactions. If we automatically disable this legitimate business traffic, that has a different effect on the organization. At the same time, there will doubtless be attacks that are not caught by our monitoring and protection devices.

Obfuscation is a technique that works both ways and we must be very careful to remember that the door swings both ways. In our rush to develop ever more robust products to catch attacks on our systems, we often forget that our enemies are doing just as much (in some cases more) to obscure their traffic so that we are unable, even with new technologies, to catch them. The fundamental nature of man is that he continues to advance his capabilities through evolving processes. Anyone who believes that they can develop an invincible technology (on either side) is guilty of a severe logic error. Any new technology that is introduced to protect, or attack resources, has an optimal lifetime associated with it during which it will be nominally effective, but given enough time, it will be defeated by other types of technologies.

Conclusions

Companies involved in electronic commerce must re-evaluate their positions with respect to where and how they conduct electronic commerce. New laws and new levels of accountability are going to require companies to take security seriously, placing it at the pinnacle of consideration when conducting electronic commerce, storing data, and when developing software. Meager sized, poorly trained security staffs, by in large, have demonstrated that they are unable to protect a company from the onslaught of attacks from well trained, well equipped, and well protected foes. Nor will minimally trained security staffs allow us to claim that we have exercised due diligence in protecting sensitive information.

Technology, which is useful in the battle against hostile elements, is merely one facet of a strategy to protect information resources, not a sufficient answer. No longer can we accept the claims of product organizations at face value, having to assume the responsibility to validate their claims through our own rigorous security testing. In addition, we must be wary of security services companies who claim that they can guarantee a secure environment challenging them to effectively put in writing their guarantees and requiring them to back up their outlandish claims with monetary resources when they are eventually shown to be in error. Instead, we need to rely on security consulting companies that truly understand the magnitude of the task before them and which can assist our organizations to evolve in our capabilities to defend against the ever evolving threats against our information and resources. Short term identification and correction of issues is at best a band aid.

Companies must adjust their thinking and realize that they are in a war. Just as some do not fully comprehend the horror of September 11th because they were not personally affected, some companies will fail to react to the cyber war in which we are engaged. In order to fight the war effectively, companies and vendors must be prepared to change the manner in which they have traditionally done business and adapt to a "security first" mindset. Companies must begin to invest in protecting their information, their customer's information, and their employee's information, taking whatever steps are necessary to do so. In addition, companies must work to develop effective Social Engineering protocols to protect key information. Although the war is in somewhat of an embryonic stage at this point, the war and the complexity of the war will escalate dramatically over the next few years as more and more rogue states and entities invest in exploiting the greatest weakness we have as a nation – our tendency towards unilateral reliance on technology. The lessons that we have learned from traditional wars, namely that our technology can not protect us from a wide variety of more mundane attacks and that our enemies can obtain access to the same technologies we have, is applicable in the information technology world as well.

Operating under the assumption that at some point in the future we will have a secure Internet environment is dangerous and we must consider carefully if it is wise to keep pouring huge amounts of resources into an un-securable environment. The constantly changing nature of the environment precludes a fully secure environment from ever being realized. Just as we upgrade our defenses, our adversaries will upgrade their offensive capability. They preserve their advantage by being able to work unnoticed on their attacks while we can only guess where, when, and what type of attack will be launched next. In many cases, they have access to the same education and the same technologies that we have and are often able to reverse engineer a technology in order to use them against us.

Our mindset must change in this area; reactive security will no longer do. Attacks have the advantage over defense in this arena, because they can be launched when individuals least expect it and because it is virtually impossible to envision and defend against every type of attack. We must realign our security posture to eliminate avenues of attack before they materialize as opposed to simply monitoring, researching, and fixing newly discovered vulnerabilities. The dynamic nature of the environment, the ever increasing complexity of the threats, and the highly active nature of the hostile elements seeking our information requires proactive security. No longer can we implement a technology and then forever patch as each new vulnerability is exposed. Nor can we assume that product companies will protect us. We must assume responsibility for our own security instead of relying on others and exercise the diligence to test technologies before we place them into production, holding vendors accountable for their lack of diligence in the same areas. We must consider a whole new cadre of strategies such as partitioning of information, division of responsibilities, usage of private networks, engaging true security experts, and re-engineering our processes internally to protect our information. We must be willing to change technologies if the vendors we currently use do not take security seriously and must begin to train our developers to be able to develop and test more secure applications.

Companies must re-think their transmission of sensitive data over public networks where hostile elements are continuously active, where transmissions are routinely intercepted and stored, and where attacks can be launched out of reach of the long arm of the law. Access is the enabler for all hostile elements to be able to attack our data. We can no longer assume that Internet usage is safe, rather we must assume that while we are using the Internet, we are constantly at risk and adapt our businesses strategies accordingly. We have taken use of the Internet for granted and are now trying to find ways to secure Cyberspace rather than consider the use of more protected networks as a way to exchange sensitive information. Methodical measurement of the ROI (Return On Investment) will allow us to determine the cost and risks associated with using a public network and allow us to determine when and where it makes sense to move much of our sensitive information transmission to a more private network structure.

While this book is not meant as a scare tactic, it is designed to present some of the sobering realities of where our world currently is and how we must adapt our business and technology strategies to be able to protect information. This author contends that the basic underlying assumption that we can truly secure Cyberspace is not realistic since there are too many human limitations, both from a political and knowledge-based standpoint, that prevent this from occurring. Even if we as technology leaders are able to raise the bar to a point where we minimize the number of attacks, there will be a very high cost to be paid, in a large number of areas, to achieve even this laudable goal.

Appendix A - Summary of the Key Security Axioms

Listed below are the key security axioms that we have stated throughout this document. The development of any layered security model should start with these axioms to prevent the wasting of resources on inadequate or ineffective security measures. It also calls for careful consideration of what conduits sensitive information should be transmitted over. Although it may appear a step backwards in electronic commerce, entities routinely exchanging sensitive information should carefully consider the utilization of private networks, managed by security experts, to exchange sensitive information.

The security axioms below, when considered, will enable us to build more protected, more resilient infrastructures that at least allow us a chance to protect our information resources.

1) Each layer of security should be implemented as though there are no other security layers

2) Just because information is protected at a point in time, does not mean that it is always protected

a) Even though information may not be used against you today does not mean that it will not be used against you in the future

3) Transmitting sensitive information, with a high liability value and a long half life, over an unsecured network, is inherently risky

4) The first step in defeating a security scheme is to know of its existence and its nature, therefore obscuring a security scheme provides some inherent protection

5) A proactive approach to security is required to effectively thwart many of the threats levied against informational resources

6) It will never be possible to fully envision and counter all possible threats against a resource in an uncontrolled environment. Therefore, in order to have a reasonable chance of protecting a resource, the sources of threats must be reduced to a manageable level

   a) It is easier to attack a resource than to defend a resource since an attacker need only find one open avenue of attack whereas the defender needs to anticipate and close all potential avenues of attack

7) A proactive approach to security is required to have any chance to thwart many of the threats levied against information resources.

8) Security companies are to be engaged with a clear goal of improving security around informational resources through current information, but not trying to guarantee it

9) No matter how much technology is leveraged to secure a resource, without the associated social engineering, it is worthless

   a) Unilateral reliance on technology to protect a resource is a recipe for disaster

10) Effective security models can not be implemented as a series of uncoordinated activities, regardless of the experience of the constituent members

11) It is cost prohibitive to try to keep up with all of the changes in security on a daily basis that an open and dynamic network requires

12) The dynamic nature of an open network requires an ongoing commitment to training for all parties involved in designing, developing, deploying, and configuring applications

13) Protect information resources as though they contribute directly to the bottom line, for in the future, they will

14) The ever evolving nature of the war requires a proactive mentality, attempting to identify and close all vulnerabilities before applications and systems are deployed

15) Effective network and information partitioning strategies inhibit attacks by reducing the target value and reducing the damage from successful attacks

16) Obscuring information and information transmission can be an effective way to protect information

Appendix B - Three Weeks of Vulnerabilities

Listed below are the software vulnerabilities identified between November 22, 2002 and December 13, 2002. Note that this is not necessarily a comprehensive list of vulnerabilities, but those that have been discovered and reported within the timeframe identified.

Vulnerability and Risk	Severity
Denial of Service vulnerability exists in the FTPD server when a malicious user sends a CEL parameter of excessive length. It may also be possible to execute arbitrary code.	Low/High (High if arbitrary code can be executed)
Vulnerability exists because the contents of shopping carts can be modified by customers, which could let a remote malicious user modify the price of items.	Medium
Multiple vulnerabilities exist: a remote Denial of Service vulnerability exists when connecting to the daemon via finger; a Denial of Service vulnerability exists when the server is caused to write to a socket that is not a listening client; and a file disclosure vulnerability exists due to the supplementary group privileges not being dropped and insufficient sanity checks of the '.plan' file, which could let a malicious user obtain sensitive information.	Low/ Medium (Medium if sensitive information can be obtained)
Vulnerability exists in the database bind() function in 'config. inc' because it is possible to bypass authentication, which could let an unauthorized remote malicious user obtain administrative privileges.	High
Denial of Service vulnerability exists when mod_ik is used due to design problems in the module.	Low
Vulnerability exists in the 'useraction. php' script due to a failure to properly check user credentials, which could let unauthorized malicious users read postings in internal forums.	Medium
Denial of Service vulnerability exists when a malicious user creates a directory, descends it, creates another directory of the same name, and then attempts to move the directory up one level	Low

Vulnerability and Risk	Severity
in the hierarchy.	
Multiple vulnerabilities exist: a Denial of Service vulnerability exists when a NULL HTTP request is submitted; and a format string vulnerability exists in the awp_log() function due to an incorrect use of the syslog() function, which could let a malicious user execute arbitrary code.	Low/High (High if arbitrary code can be executed)
Buffer overflow vulnerability exists in the 'index. cgi' script when a parameter is submitted that is of excessive length, which could let a remote malicious user execute arbitrary code.	High
Buffer overflow vulnerability exists due to a lack of validation of requests, which could let a malicious user execute arbitrary code with 'bin' level privileges. *Note: Canna is typically installed only when Japanese language support is enabled.*	High
Buffer overflow vulnerability exists prior to authentication because overly long strings are not properly handled, which could let a remote malicious user execute arbitrary code.	High
Denial of Service vulnerability exists in Optical Service Module (OSM) Line Cards when an irregularly constructed network packet is processed.	Low
Vulnerability exists when a specially crafted command line argument is submitted, which could let a malicious user, obtain elevated privileges and execute arbitrary programs.	High
Vulnerability exists when the system is configured with the incremental scan option, which could allow virus or Trojan code to be saved to disk, and signed as clean.	Medium
Several buffer overflow vulnerabilities exist: a vulnerability exists in the SASL Library due to insufficient bounds checking while sanitizing usernames, which could let a malicious user execute arbitrary instructions; (Note: This issue only exists if the default realm is set.) a vulnerability exists in the SASL library due to a failure to allocate sufficient memory when it is required to escape characters, which could let a malicious user execute arbitrary code; and a vulnerability exists when log files are generated due to a failure to allocate sufficient memory, which could let a malicious user corrupt memory	Medium/ High (High if arbitrary code can be executed)

Vulnerability and Risk	Severity
and possibly cause inaccurate logs to be created.	
Buffer overflow vulnerability exists in the Sieve library when a header of excessive length is submitted, which could let a remote malicious user execute arbitrary code.	High
Several buffer overflow vulnerabilities exist: a buffer overflow vulnerability exist in the Sieve library when an IMAP flag of excessive length is passed to the program, which could let a remote malicious user execute arbitrary code; and a buffer overflow vulnerability exists when excessive error messages are generated, which could let a remote malicious user execute arbitrary code.	High
Vulnerability exists due to the way Debian Internet Message (IM) creates temporary files, which could let a malicious user corrupt or modify data.	Medium
Vulnerability exists in the OPTIONS directory when attempting to handle a malformed request for a resource, which could let a malicious user cause a Denial of Service and possibly execute arbitrary code.	Low/High (High if arbitrary code can be executed)
Vulnerability exists because temporary files are created in an insecure manner, which could let a malicious user obtain elevated privileges.	Medium
Vulnerability exists when the WINDOWS + F key combination is held down for an extended period of time, which could let a malicious user bypass security restrictions and obtain unauthorized access.	Medium
Vulnerability exists due to inadequate input checks when a NLST response is received from an FTP server, which could let a remote malicious user overwrite files on the client system.	Medium
Buffer overflow vulnerability exists in GNUPlot that is shipped with SuSE Linux, which could let a malicious user obtain root privileges.	High
Vulnerability exists in 'rwords' filtering, which could let a malicious user bypass e-mail filters.	Medium
Several buffer overflow vulnerabilities exist due to insufficient bounds checking, which could let a malicious user, cause a Denial of Service and potentially execute arbitrary code.	Low/High (High if arbitrary code can be executed)
Vulnerability exists in the ied program, which could let a malicious user obtain sensitive information.	Medium
Denial of Service vulnerability exists in the xntpd	Low

Vulnerability and Risk	Severity
program.	
Vulnerability exists because the installation of HP-UX Visualize Conference may leave certain directories with insecure permissions, which could let a malicious user obtain unauthorized access and elevated privileges.	Medium
Vulnerability exists because HTML is not properly sanitized from user profile photo URIs, which could let a malicious user execute arbitrary code.	High
Vulnerability exists in the X-Forwarded-For: HTTP header proxy fields because HTML is not properly sanitized when the header field is logged, which could let a malicious user execute arbitrary script code.	High
Vulnerability exists in the KIO subsystem rlogin and telnet protocols, which could let a remote malicious user execute arbitrary commands.	High
Vulnerability exists when the Apple Package Manager is used to install the application because file permissions are changed to unsafe modes, which could let a malicious user obtain sensitive information.	Medium
Directory Traversal vulnerability exists, which could let a remote malicious user obtain sensitive information.	Medium
Information disclosure vulnerability exists in SuidPerl, which could let a malicious user obtain sensitive information.	Medium
Vulnerability exists in some default configurations because data held in third-party relational databases is stored insecurely, which could let an unauthorized malicious user obtain sensitive information.	Medium
Several vulnerabilities exist: a buffer overflow vulnerability exists due to insufficient allocation of memory, which could let a malicious user execute arbitrary code; a vulnerability exists because no authentication is required to access a xml page, which could let a remote malicious user obtain unauthorized access; a buffer overflow vulnerability exists due to insufficient allocation of space for local buffers, which could let a malicious user change configuration information on the vulnerable device; and a Denial of Service vulnerability exists due to insufficient bounds	Medium High (High if arbitrary code can be executed)

Vulnerability and Risk	Severity
checking when copying user-supplied input to memory.	
Denial of Service vulnerability exists in the XML parser that is used by these products.	Low
Multiple vulnerabilities exist: a vulnerability exists because it's possible for an un-trusted Java applet to access COM objects, which could let a malicious user obtain control over the machine; two vulnerabilities exists because it is possible to spoof the location specified in CODEBASE parameter in the APPLET tag, which could let a malicious user obtain sensitive information; a vulnerability exists due to a flaw in the Virtual Machine's URL parser, which could let a malicious user intercept any traffic that the user would send to the trusted site; a vulnerability exists because Java Database Connectivity APIs don't properly regulate who can call them, which could let a malicious user obtain sensitive information; a Denial of Service vulnerability exists due to insufficient security checks in the VM; a vulnerability exists because VM doesn't prevent untrusted applets from accessing the user. dir system property, which could let a malicious user sensitive information; and a vulnerability exists because it is possible for a Java applet to create an incorrectly initialized Java object, which could let a malicious user cause Internet Explorer to fail.	Low/Medium/High (Medium if sensitive information can be obtained and High if control can be obtained over the system)
Information disclosure vulnerability exists in the wireless LAN feature, which could let a malicious user obtain sensitive information.	Medium
Vulnerability exists because it is possible to bypass Internet Explorer's cross-domain security model when using object caching in scripting flaw due to incomplete security checks, which could let a malicious user execute arbitrary code.	High
Vulnerability exists in the showModalDialog and ShowModelessDdialog functions when script code is injected into the style parameters due to improper checks, which could let a remote malicious user execute arbitrary JavaScript and HTML code	High
Denial of Service vulnerability exists if an e-mail message is submitted that contains a partially malformed header.	Low
Vulnerability exists because it's possible for one	High

228

Vulnerability and Risk	Severity
process in the interactive desktop to use a WM_TIMER message to cause another process to execute a callback function at the address of its choice, even if the second process did not set a timer, which could let a malicious user obtain full administrative privileges.	
Design error exists in the Win32 API inter-window message passing system, which could let a malicious user obtain elevated privileges.	Medium
Vulnerability exists in the Fast User Switching (FUS) option that allows users that have been downgraded from the Administrator to a normal user to still use the Task Manager, which could let a malicious user obtain sensitive information.	Medium
Vulnerability exists in the negotiation process because it is possible to cause the signing of Server Message Block (SMB) packets to be disabled, even when it is required by the host, which could let a malicious user obtain sensitive information.	Medium
Buffer overflow vulnerability exists when malformed POST requests are submitted, which could allow a malicious user to cause a Denial of Service.	Low
Directory Traversal vulnerability exists due to improper sanitization of web requests, which could let a remote malicious user obtain sensitive information.	Medium
Buffer overflow vulnerability exists when an overly long value is submitted for the FTP change directory (CD) command, which could let a remote malicious user execute arbitrary code.	High
Vulnerability exists because several FTP clients distributed with various operating systems may handle NLST FTP responses in an insecure manner, which could let a malicious server overwrite key files to cause a Denial of Service or, in some cases, gain privileges by modifying executable files.	Medium
Buffer overflow vulnerability exists in the XFS font server, fs. auto used by multiple vendors, which could let a remote malicious user execute arbitrary commands.	High
Multiple vulnerabilities exist: a Denial of Service vulnerability exists due to the way the Apache	Low/High (High if arbitrary

229

Vulnerability and Risk	Severity
scorecard is handled; a Cross-Site Scripting vulnerability exists due to improper sanitization of SSI error pages, which could let a malicious user execute arbitrary HTML or JavaScript code; and a buffer overflow vulnerability exists in the ab. c web benchmarking support utility, which could let a malicious user execute arbitrary code.	code can be executed)
Vulnerability exists in 'dvips' when a maliciously constructed file is passed to the lpd daemon, which could let a malicious user execute arbitrary commands.	High
Directory Traversal vulnerability exists due to a failure to properly sanitize web requests, which could let a remote malicious user obtain sensitive information.	Medium
Directory Traversal vulnerability exists due to a failure to properly sanitize web requests, which could let al malicious user obtain sensitive information.	Medium
Vulnerability exists in the IP Queuing module due to insufficient checking of the integrity of the privileged process, which could let a malicious user obtain sensitive information.	Medium

Source: National Infrastructure Protection Center December 16, 2002 Report

Appendix C - Survey of Major Internet Attacks

There have been a large number of attacks against different types of entities. Attacks have accelerated over the past few years, which can be expected since more and more entities are using the Internet for more and more things. This is a very small list. Since most companies in the past have not been forced to report all of their intrusions, it is unclear how many hundreds or thousands of additional attacks have occurred. A few of the more notable attacks are listed below.

Date of Attack	Nature of Attack	Target
February 2003	Personal information of employees stolen	FTD
February 2003	8,000,000 credit card numbers stolen	Third party credit card processor
January 2003	Attack on Microsoft SQL Server	Microsoft SQL Server sites
October 2002	Denial of service attack resulting in slower resolution times	Root DNS servers
June 2002	$31,000 siphoned from various bank accounts	Singaporean Bank DBS
April 2002	3,600 bank customers records were stolen	Republic Bank
April 2002	265,000 personal records, including social security numbers were stolen	State of California
March 2002	60,000 personal records stolen	Prudential Insurance

Date of Attack	Nature of Attack	Target
February 2002	Bid for $158,000,000 in US Securities without paying for them	US Treasury
June 2001	Access to banks systems for more than six months	Central Texas Bank
March 2001	Stolen credit card numbers (identified as the largest attack to date)	More than 40 Internet Banking and Commerce sites
May 2000	Love bug virus does an estimated $10 billion in damages	Many sites worldwide
March 2000	28,000 credit card numbers stolen with losses approximated to be $3.5 million	Multiple web sites
February 2000	Denial of service attack costing "millions of dollars"	Yahoo
March 1999	Extortion based upon ability to break into systems	Bloomberg

About the Author

Gregg Powers, an Information Technologist, has more than 25 years experience with all facets of Information Technology. He has directed architecture, as well as research and development activities, in a number of different large organizations and provided electronic commerce consulting to a variety of organizations through his own consulting company. He has provided consulting to the State of Colorado Governor's Office of Innovation and Technology on security, electronic commerce, and related subject matter areas. Gregg is currently the Chief Technology Officer for one of the fastest growing county governments in the nation. Gregg has his Bachelors of Science in Computer Science from the University of Colorado.

www.ingramcontent.com/pod-product-compliance
Lightning Source LLC
Chambersburg PA
CBHW051231050326
40689CB00007B/881